LARGE VOLUME WATER DELIVERY

LARGE VOLUME

WATER DELIVERY

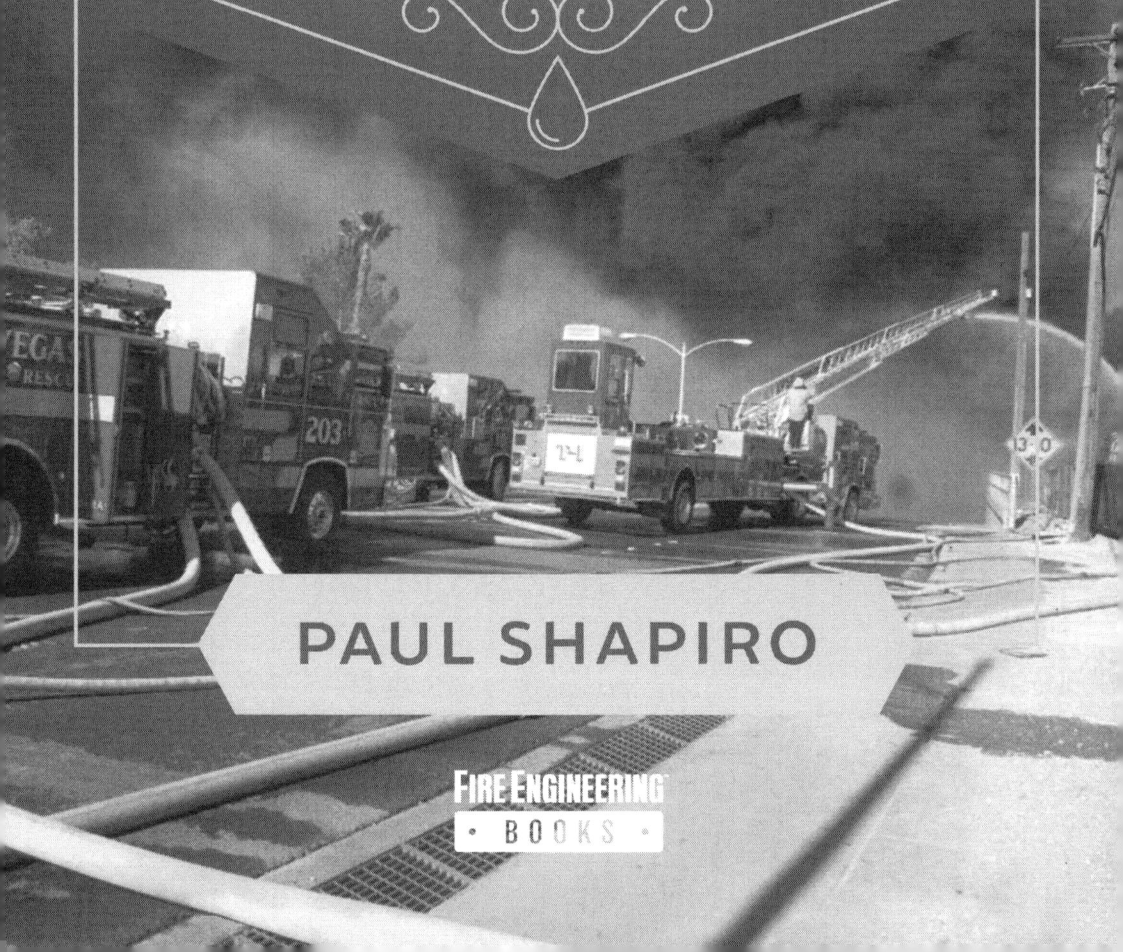

PAUL SHAPIRO

FIRE ENGINEERING BOOKS

Fire Engineering Books
110 S. Hartford Ave., Suite 200
Tulsa, Oklahoma 74120 USA

800.752.9764
+1.918.831.9421
info@fireengineeringbooks.com
www.FireEngineeringBooks.com

Senior Vice President: Eric Schlett
Operations Manager: Holly Fournier
Sales Manager: Joshua Neal
Managing Editor: Mark Haugh
Production Manager: Tony Quinn
Developmental Editor: Chris Barton
Cover Designer: Jared Hood
Book Designer: Robert Kern, TIPS Technical Publishing, Inc.

Library of Congress Control Number: 2020944398

Printed in the United States of America

2 3 4 5 29 28 27 26 25

This book is dedicated to my father, Bennie Shapiro, who was part of the Greatest Generation and fought in the European theater during WWII. Dad passed away when I was 12 years old, but in those 12 years he gave me an education about life that has lasted a lifetime. Dad worked six days a week and sometimes two or three jobs to support his wife and three children and make life as good as possible for his family. He always had time for us no matter how tired he was.

Dad, you've been gone a long time, yet I feel as though it was yesterday having you as my father. Dad, this one's for you. I love you with all my heart.

Your son,

Paulie

CONTENTS

II. DISCHARGE WATER DELIVERY

PREFACE

This book is rather unique among fire service texts because it goes into detail concerning how to accomplish specific tasks involved with large-flow water delivery operations. The information presented in this book was derived from actual flow tests with certified flow testing equipment, as well as from hands-on testing to ensure the information presented was accurate, safe, and efficient on large fires. I consulted with manufacturers of water delivery equipment to make sure that the equipment being used falls within the guidelines set for them.

This book is based on actual fireground experience rather than standard operation procedures, assumptions, and habits. Large-flow fires are unique for several reasons. Most agencies do not frequently experience them, so their hands-on experience in coping with water delivery for a fire of this magnitude is low. As a result, there may be a lack of appropriate policy, an absence of equipment needed to deliver the required flow in a safe and efficient manner, and minimal training in the necessary multicompany operations.

My recommendation is to use this book as a reference when conducting training and introducing new concepts and equipment that may better suit your department. We must always be looking for a better way to do things.

ACKNOWLEDGMENTS

I would like to thank John Ryan, who was Chief of Training with the Las Vegas Fire & Rescue Department way back when. He believed in me and gave me a chance to develop a new water delivery system for our department early in my career. It was this project that introduced me to the world of water delivery and has allowed me to continue ever since. Thank you, Chief.

I would like to thank my wife, Nancy, for her endless help in putting together this book as well as helping me run my consulting business, Fire Flow Technology. Thank you Nancy, I love you.

INTRODUCTION

Water has been used as a fire extinguishing agent for most fire extinguishment efforts throughout the years. Some fires require specific agents, but for the everyday fire attack experienced all across the globe, water is the number one extinguishment agent.

From a nonfirefighter perspective, it may appear that a firefighting operation is all about lobbing water onto a fire any way possible, and no art or technique is involved. We often have someone watching us in action and criticizing what we do. The fact is there are many tactics and strategies that go into developing adequate fire streams to match the heat energy (denoted in British thermal units or Btus) generated by the fire.

I started in the fire service in 1981 when training was a bit simpler. There were two major topics to be learned in rookie school—emergency medical service (EMS) and good old-fashioned firefighting. The EMS portion involved anatomy, bandaging, patient assessment, CPR, etc. On the firefighting side we were taught the basics (I do mean basics) of all the physical aspects of fighting the fire. This included everything from riding tail board (yes, I did) to pulling hose, laying supply lines, forcible entry, hose handling, and a thorough understanding of our weapon—the nozzle. What we were not taught was the fire science portion of water application and how much water it takes to put out a specific size of fire; the water flow rate (gpm) versus the fire-generated heat energy (Btus). So in an interior attack we would pull the 1¾" line, and for the bigger outside fires (defensive fires), the 2½" line or master stream appliance was the weapon of choice. Again, there was no training on the topic of matching the water flow to the heat generated by the fire.

Fast forward 35 years and take a look at the current rookie schools. It seems like they are teaching a little bit of everything: EMS, firefighting, hazmat, swift water rescue, how to be nice to one another, etc. This change in training emphasis has affected the fire service. One change is that there is a bigger training workload. As a result, sometimes only the basics of many different topics are addressed, without getting into the meat of the issues. When that is the case all topics suffer, including firefighting. Some might argue that firefighters get all their training on the job because they get called to numerous fires. You may respond to a lot of fires, but if you are doing the wrong thing at these fires, are you really learning?

This book focuses on how to match the water flow to the heat energy generated. This is addressed for one main reason: to get a quick knockdown on the fire. This book is meant for all personnel involved with fire suppression operations from the rookie and veteran firefighters to the pump operators, company officers, incident commanders, and the training officers who will spread the word.

So how do firefighters achieve the level of knowledge at which they are ready for the fight? First, we all need to be on the same page about what to do and how to do it. Next, we need to train effectively, not for the sake of completing mandatory hours required by the department, but to learn how and why we conduct our water delivery tactics. This involves everything needed to produce the required stream from water supply to discharge evolutions, followed up with proper nozzle placement. This training should be ongoing. Special emphasis should be directed toward events that rarely occur, such as high-rise fires and multicompany operations. Make the training as realistic as possible. Have simulated fire problems with units actually responding. If master streams are involved, aim the stream as though you are hitting the fire, not the sky.

This book will also focus on the mechanics of moving water, with the addition of hose handling techniques and information on tactics and strategy. The philosophy of this book is to introduce tactics that are progressive, as safe and efficient as possible, and practical in real-world situations. I will be discussing information that is not in the standard fire stream books being used today. In fact, some of the things I recommend might even go against current guidelines. I spent numerous hours researching and flow testing and developing many of the tactics here. My guidelines for the development of the specific use of equipment is simple. I always follow the rules and specifications for all of the equipment being used and always keep safety as the priority. This is all about flowing large volumes of water for large fires, which in itself has the potential to make for safer operations on the fireground (fig I–1).

The citizens of your community deserve the best you can give them. As long as the task (in this case, water delivery) is feasible, even if some of it is a little complicated, all students should be brought up to the level needed to accomplish it. Too often I see departments lowering the task requirements to fit the average firefighter. Treat this education as though the skills learned are going to be used to protect your family. Average should not be acceptable.

My Motto: Where There's a Will, There's a Way

Throughout the years, fire instructors and fire departments have laid the groundwork for water delivery on the fireground. Some of the procedures instituted are still in use today and remain effective. On the other hand, some of the

Figure I–1. Where there's a will, there's a way. This unit is pumping a 4" discharge line from its deck gun.

Photo courtesy of Alan Simmons Productions.

procedures are now outdated or impractical. Newer procedures and technically improved equipment allow us to do our jobs better. This is a good thing because staying static ensures that nothing will change. I am one of those guys who has offered my two cents' worth, and I feel like I have developed some good ideas for tactics and equipment. My first guideline to follow is to never go against factory recommendations set forth by the manufacturers of the equipment being used. My second guideline is to be safe and confident in the use of the equipment and tactics, whether with a higher flow and pressure on a smoothbore tip or with developing looped supply line evolutions. I based my studies on the following eight rules.

Big Paulie's Words of Wisdom on Moving Big Water

Rule 1. Tradition and pride can be hazardous to your health.

Ah yes, tradition and pride—they are the mainstays of the fire service. The idea is fun. It promotes camaraderie and is easy to attain. It is embodied, however, by a mindset that may seem familiar:[1]

- "We've always done it this way, and the fires have always gone out."
- "We are an aggressive fire department; we go in and attack the fire."
- "Fire departments that make exterior attacks are known as 'foundation savers.' They are a bunch of wimps."
- "Pump 150 psi."
- "Smoothbore tip nozzles are the only ones that work."
- "Automatic nozzles are the only ones that work."
- "We don't get those kinds of fires here, the large ones." (This is my personal favorite.)

The list goes on, and I'm sure you have your favorites. If you analyze each of these statements, you can find one common denominator: it's easy. You do not have to change anything, there are no studies or evaluations to do, and there is no risk-taking to endure. You just do the same thing over and over again. This fails to address the fact that the fire service is constantly changing. We have gone from horse-drawn quick attack units to motorized apparatus, and we have traded in our buckets for hose. We have gone from no air packs to using air packs to filter out dangerous air pollutants. And of course, we have graduated from 2½" supply lines to large diameter hose. I would like to have been a fly on the firehouse wall when some of these changes were made, or for that matter, just suggested. Are you starting to get the picture here?

Big Paulie's first recommendation is to *please* keep an open mind. Tradition and pride are important to the fire service, but please do not let them get in the way of making our job easier and, more importantly, safer.

Rule 2. Keep it simple.

Firefighting definitely gets high marks as far as the degree of difficulty is concerned, both mentally and physically. We deal with life-threatening situations that we did not create, and we do not get any say in when and where they will happen. Therefore, I feel it is extremely important to make our job as easy to perform as possible. What does this mean when it comes to flowing big water? Let's take a look.

Get rid of the formulas. We have all been taught the formulas for calculating the required fire flow for a structure fire. First of all, who says that one person's approximation of what the dimensions of a structure are will be the same, or even close, to what mine or the next person's will be? Second, how sharp are you going to be in your estimation at 2:00 a.m. when you are first in and the whole world seems to be on fire? If you are like most of us, things could be a little difficult to discern precisely. Instead of using this formula, why not break fire flows down into three categories: small, medium, and large?

Small fires can be, for lack of better words, the junk fire flows that require small amounts of water to extinguish (50 gpm should work well). You know these types of fires, small trash or brush fires, or possibly a smoldering fire under the hood of a vehicle. We typically pull the booster line on these types of fires.

Medium fires are for the most part the interior compartmented structure fires in which an interior attack will be attempted. Flows of 120 gpm to 150 gpm seem to work well on these fires.

Finally, *large fires* are basically just that—the big ones. They can be large structures where both exterior and/or interior are fully involved, large vehicle fires, etc. Basically, these fires encompass anything that requires more than the standard small diameter handline to extinguish rapidly. What kind of flows are we looking at? We need as much as possible, given the available water supply, manpower, and equipment. I would not hesitate to hit a fully involved two-car garage fire with a 1,000 gpm stream from a deck gun under the right circumstances. The bottom line is to hit the fire with as much water as you can develop based on the circumstances that are present.

Calculating engine pressures. Fireground hydraulics can be complicated. Books are written on the subject, and you can take college courses devoted to this topic. Consider the following scenario. It's noon, and you are first in on a *ripper* (working fire). The firefighters are screaming for water. The first evolution deployed is at least 200' of 3" hose, attached via a wye to a manifold with two 1¾" handlines and one 2½" handline. The requested flows are as follows: 1¾", 150 gpm each; 2½", 250 gpm.

Now quickly, what is your engine pressure? Come on, hurry up, we need the water! We are getting clobbered here!

I think hydraulic formulas have no place on the fireground. They are too difficult and require all of the figures to use. What if you do not know how much hose is on the ground? The formulas will not work. One of my favorite formulas that I really get a kick out of is the one for relay pumping. You are asked to establish an engine pressure for the source pumper based on the size and length of the hose and flow requirements. *True fact*: in a relay pump operation, which could be long, the amount of hose is usually not known. Sometimes it takes two or more units to complete the lay. How about the flows? Most of the time the actual flow is not known until the evolution is completely set up and engine pressures are established.

Let's keep it simple. Pump charts can and should be used to develop all engine pressures. Pump charts are reference charts that have precalculated pressures for any hose evolution your department will ever have to perform. The more exact engine pressures that can be listed, the better. For example, pump charts could include preconnected handlines, fixed master streams, foam operations, sprinkler operations, high-rise building fire protection systems, and relay pumping.

For some of the make-or-break evolutions, basic math should be all that is needed. A listing of statistics will help this process, such as the friction loss per 50' of hose for various sizes, nozzle pressures, master stream appliance friction losses, elevation gain or loss, etc.

To make a long story short, always be looking for the simplest way to accomplish a task. Can the current method be done at 2:00 a.m. under intense stress and with minimum manpower? If not, then it is time to reevaluate and find an easier way.

Rule 3. The water's out there. You just have to go get it!

How often have you heard the excuse of a poor water supply as the reason for losing a building to fire? Sometimes there are evening news reports about major fires where everything was lost. In the accompanying interview, the fire chief puts the blame on a poor water supply. Yet in the background, you can see large diameter hose on the ground, and the fire building is in the middle of a major city. Major cities and even average-sized cities for the most part have hydrants available for miles. How then does the water supply problem occur? What usually happens at big fires is that the closest hydrants to the fire building are connected to supply lines because they are the easiest ones to access. As the fire gets bigger, the water supply in the immediate area of the fire building diminishes. This is where most departments, especially the ones using large diameter hose, give up and say they are out of water. They think if they cannot get water with a 5" line, there must not be any more water available. In reality, they are not out of water in terms of being able to obtain more. The water is out there; they just need to go get it. Long-distance relay supply line operations and water shuttle operations are two ways to obtain more water. For a big flow operation, preplanning of the potential high fire flow areas in conjunction with careful appraisal of the department's available equipment will be essential to developing an effective plan.

It is true that there are situations where genuine water supply problems exist. Before throwing in the towel, however, look at all your options. You never know when you might come up with a solution.

Rule 4. Don't screw around; just blast it!

"Elephant hunting with a BB gun" is what I like to call an attack on a big fire with a low-flow stream. There is a basic rule of thumb given to us by Mother Nature that says a certain amount of water will be required to put out a certain amount of fire. This is the plain and simple truth. If the amount of water is insufficient to put out the size of fire you are dealing with, the fire won't go out (at least not quickly).

There are formulas that tell us how much water is needed for a specific size of building. This is good for books, but not good for practical application. The bottom line is to hit the fire with as much water as possible based on your manpower, equipment, and water supply situation in an attempt to overwhelm the fire. If 250 gpm will knock a fire down fairly quickly, then 500 gpm will do so even faster.

Take a look at a common scenario that cities deal with, a mobile home fire. In this scenario there is a full assignment responding to the fire. The first-in engine company has laid a 5" supply line from the hydrant located about 300' away. They have 500 gallons directly available from their booster tank and are faced with two immediately threatened exposures, one on each side of the fully involved mobile home. What usually happens is that the crew will pull preconnected handlines to cover the exposures until the water supply can be established.

The key here is to protect exposures while conserving the water from the booster tank. With two lines out, you know as well as I do that there is a chance of draining the 500 gallon booster tank before the supply line is charged. Furthermore, the fire that was creating the exposure problem is still there but is

probably burning with more intensity, and now we are out of water. Have we conserved water, or wasted it? Use the same scenario again but change the initial attack. We are still going to lay the supply line, but this time instead of pulling the preconnected handlines, we are going to get on top of the engine, swing that deck gun around, and blast the fire building with everything we have. Our goal is to quickly overwhelm and extinguish the fire. Even if we do not knock it down completely, there is a very good chance that we will transform it from an raging inferno into a lazy type of fire that is going nowhere quickly. This should take only 15–20 seconds. At a flow rate of 500 gpm, a 20-second blast will use only 166 gallons of water. This type of attack is known as a *blitz attack*.

Rule 5. The only thing too much water will do is put the fire out faster.

If it takes 30 seconds for a 200 gpm stream to extinguish a fire in a building, then a 500 gpm stream will obviously need less time to extinguish the same fire. This is a simple concept. Do not be afraid to deliver as much water as possible to put the fire out. The key to this operation is to overwhelm a fire, stopping it as quickly as possible, and then shut the stream down. If done correctly, there is a good chance the fire will be extinguished more quickly and with less water.

Rule 6. It's easier to design a pumper to flow large volumes of water and not need it than it is to flow large volumes of water from a pumper not capable of doing it.

All pumpers should be designed for flowing large volumes of water. This involves big pumps, large plumbing, large master stream appliances, multiple large intakes, and large diameter hose (the larger, the better). What this does is decrease the friction loss in all parts of the water delivery system, which in turn decreases the horsepower needed from the engine of the pumper. All this makes for a more efficient operation. The chances for doubling a pump's capacity with a unit designed for big flows are good. It will easily handle the job when small flow are needed as well. If you are not yet convinced, consider the alternative. How easy will it be for a unit designed with standard plumbing, appliances, and hose to push the big water?

Rule 7. Pump it up!

Here is another one of those Mother Nature rules. Water moving through fire hose, nozzles, and discharge plumbing requires a positive pressure. High-flowing and hard-hitting streams can require higher-than-normal engine pressures to produce. Sometimes in the fire service we can be paranoid around these higher pressures for fear of equipment failure and potential accidents. There is no need to worry, however, since the manufacturers of our fine equipment also have that fear. That is why they have designed their equipment to withstand the pressures needed for the big water operations.

Keep in mind that all water passage components within the fire pump system of a fire apparatus have to be able to withstand 600 psi. Most small diameter handline hoses today come with a 400 psi service test pressure. According to NFPA standards, this pressure needs to be 10% higher than maximum operating pressure.[2] That gives us a maximum working pressure of 360 psi.

Manufacturers of master stream appliances have given us a maximum working pressure ranging from 175 to 200 psi depending on the make of the appliance. Keep in mind that when these pressures were established, the manufacturers knew that firefighters would be using the equipment and could possibly mess up and overpump with it. Thus a safety factor was built in. The above-mentioned equipment can actually withstand more than the maximum pressures they have given.

As a result, 1¾" handlines can produce 300 gpm streams, 2½" and 3" handlines can produce 600 to 800 gpm streams, and smoothbore tips on master stream appliances can operate at nozzle pressures reaching 175 psi, capable of producing "killer" streams.

Rule 8. Unless it's written by the manufacturer, it can be challenged.

Training in the fire service is provided by hand-me-down information from fire service textbooks, fire department operational guidelines, and fire department instructors. The fire service is constantly changing. We need to be able to adapt to these changes and improve the way we do things. For some reason the fire service prefers to avoid change when it comes to water delivery, as is evident from some of the equipment and procedures still in use. We need to take a hard look at how we currently do things and see if there is a need for improvement. Do not be afraid to invent new procedures and techniques that may enable us to do our jobs better. Just because it is not written yet does not mean it cannot be done. Be brave enough to be the pioneer. The only limiting factors in our quest for improvement should be the manufacturer's guidelines that are established on the equipment that is being used.

ABOUT THE AUTHOR

Paul Shapiro specializes in the research, development, and training of large flow water delivery systems and fire stream management. His extensive research with the use of large diameter hose as both supply and discharge lines has been published frequently in fire service trade magazines. Paul has made significant contributions to fire response efficiency with his application-based research with large-caliber fire streams and handline and master stream operations.

Paul has been involved with the fire service since 1981 and served Las Vegas as an engineer with the City of Las Vegas Fire & Rescue for 28 years until his retirement. The department honored him as Engineer of the Year for 2006. He is a certified Fire Instructor III for the State of Nevada, has served on the faculty of many fire academies throughout the United States, and was named Instructor of the Year in 1999 by the Colorado Fire Academy. Paul is recognized both nationally and internationally as an expert in his field. Take a look at Paul's website, fireflowtechnology.com, for more information.

1

WATER SUPPLY

1

WATER SUPPLY FUNDAMENTALS

Introduction

The fire service has made two major changes to water supply delivery since first using fire hoses to deliver water from the source to the pumper.

The first major change was to incorporate pressurized water sources with hydrant systems and their supporting infrastructure, utilizing the ability to take 2½" hose supply lines from a hydrant to the pumper on the fireground (fig. 1–1). This system lasted for decades, with some significant changes to tactics along the way. Developments in techniques and equipment have included the introduction of 3" hose, deploying dual lines, relay pumping from the hydrant, and the adoption of new equipment to improve supply line operational efficiency.

The second major change was the introduction of large diameter hose (LDH). Due to its greater diameter, the larger hose greatly reduced friction loss in the hose evolution, thus allowing hydrant system pressures to move water at a higher volume

Figure 1–1. Two outlets vs. one makes sense for big water.

and for longer distances. The introduction of large diameter hose resulted in noticeable improvements in water delivery compared to the 2½" and 3" hose. The early use of large diameter hose gave rise to common sayings such as, "Laying a large diameter hose supply line is like taking the hydrant right to the fire," and "Laying a large diameter hose is like laying an aboveground water main." Firefighters quickly appreciated the benefits that the large diameter hose offered due to its ability to deliver greater volumes of water in less time than the smaller diameter hose.

Before large diameter hose was incorporated into use by the fire service, a fire department facing a large flow fire had to stretch multiple supply lines. Most of these lines were relay pumped from a hydrant. Additional supply lines would be laid, and the relay operations would keep increasing until the water demands were met. Before the use of large diameter hose, it was common to see piles of hose coiled in the street at the fire scene. Most of these piles consisted of supply lines.

Large diameter hose has been available for several decades, and most departments use it. Nevertheless, water supply issues still occur at major fires that demand large volumes of water. The reason for this is that some firefighters are treating the hose like it is a miracle hose. They lay the line from the hydrant without regard to the distances involved and still expect big water output. After all, it usually works on house fires. The true fact is that large diameter hose, as good as it is, is not a cure-all for large flow fires. The hose evolution operations that were used prior to the introduction of large diameter hose still need to be conducted.

This section will cover supply line evolutions currently in use throughout the United States, along with some new practices. A few of the tactics may be new to you, but I can assure you that they have been tried and proven, and they fall in line with all manufacturers' specifications for the corresponding equipment.

Static and Pressurized Water Sources

It is necessary to have a system in place to allow water to be used quickly when the fire department responds. There are two types of water supplies that we count on in the fire service—*static water sources* and *pressurized water sources.*

Static water sources

Static water sources come from either natural or man-made storage areas. *Natural storage areas* include bodies of water such as lakes, rivers, and even the oceans (fig. 1–2). *Man-made static water sources* include portable folding tank water tenders and booster tanks on the specific fire units. There are several methods that can be used to obtain water from natural or man-made water sources. Drafting with a water pump, whether from a natural body of water or from a portable folding tank, is the most common method. When using a water pump, it is important to remember that its output is limited by its rated capacity when drafting, and pressure is limited to 150 psi during operations.

Figure 1–2. The Valdivia Fire Department in Chile takes advantage of a river for a water supply source.

Water tenders can deliver water in different ways. They can use *head pressure*, which relates to the height of a liquid column and the pressure it produces at the base of the column,[1] by dumping the water in the folding tanks. Alternately, some water tenders have pumps and are able to deliver the water through a pump operation. One thing to remember about onboard tanks, whether from water tenders or engine companies, is that water is directed to the pump through plumbing commonly called *tank-to-pump plumbing*. The plumbing is connected to the bottom of the tank, and the pump is situated below the tank, which allows the head pressure from the tank to move the water through the tank-to-pump plumbing into the pump, unless specified differently. Per NFPA specifications, the maximum allowed water rate from a booster tank to the pump through the tank-to-pump plumbing is 500 gpm.[2] Some units can have larger capabilities, which must be specified in the blueprints of the unit being built. This book will focus on pressurized water supply systems, although much of this information also applies to static water supply operations.

Pressurized water sources

A *pressurized water source* as it relates to moving water on the fireground can be defined as a water system that starts from a static water source and is moved under pressure through water mains, with fire hydrants branching off of them. The pressure from the static source can be head pressure, pump pressure, or a combination of both. The fire hydrants are used for fire department pumpers to make the connection from the pressurized water source to fire department pumpers. There are several ways to connect to the fire hydrant in order to supply pumpers on the fireground, all of which require the use of *supply lines*. Supply lines are fire hoses used to move the water supply to the fire department pumpers. With most small flow fires, firefighters will connect the supply hose directly to the hydrant and lay a supply line to the fire, allowing hydrant pressure alone to move the water through the supply hose to the pumper. For maximum flow or when a hydrant is lacking pressure, pumpers can be placed at the hydrant to boost the pressure through the supply hose to deliver more water (fig. 1–3).

One important fact to understand about fire hydrants relates to their pressure ratings. There are two ratings for hydrants when connected to hose—*static pressure* and *residual pressure*. The static pressure shows the potential pressure the hydrant system has for moving water. Remember, however, that static pressure is the pressure of the water more or less at rest other than normal city usage of the water. In other words, water is not flowing from the hydrant itself. So if someone tells you that they have a hydrant rated at 90 psi, and thus it will be useful for big flow operations, you should realize that may not be a valid assumption. It is the residual pressure of the hydrant that moves the water, and it is based on water volume in the main.

It may be helpful to consider several examples of this principle. Let's first consider a hydrant system with 90 psi static pressure. After flowing 1,200 gpm, it still has a pressure of 60 psi. This tells me that this particular hydrant system

Figure 1–3. This old Crown is set up in a dual 5" relay.

is a high-volume system. A second example stems from events I have actually seen during training. The hydrant system had a 150 psi static pressure. After flowing the hydrant, we were only able to get 400 gpm before we ran out of water. This was a high-pressure, low-volume hydrant system. In fact, it was only fed by a 3" main. The third example involves a 60 psi static pressure hydrant system. After flowing 1,000 gpm, the residual pressure was 50 psi. This hydrant did not have adequate pressure to move the water through the supply hose. Placing a pumper at the hydrant to build the pressure worked well, however. For the most part, we get the hydrants we get at a real fire, at least initially, and we have to make do with whatever performance is available. As you will see later in this book, there are ways to utilize secondary hydrants to support the initial evolution started off of the first supply line.

Back to Basics: Deploying a Large Diameter Hose Supply Line

You are probably wondering what there is to know about laying a large diameter hose supply line. We are already aware of the mindset that this big hose is a magical solution. I hope to prove to you that there is more to the use of large diameter hose than just laying it out (fig. 1–4). It must be used properly in conjunction with all other equipment, manpower, and the local water system to achieve maximum performance.[3]

As simple as it may seem, there are several important aspects of supply line evolution deployment to ensure smooth and efficient operations.

Taking the hydrant using a hydrant strap

Let's start off with wrapping the hydrant. There are several techniques that are taught and used throughout the fire service on how to best anchor the supply line to the fire hydrant to allow the engine company to pull away, laying the line. For the most part these techniques are geared toward safety. If for some reason the supply line were to get hung up and pull away from the hydrant wrap, it is very important for the firefighter taking the hydrant to be standing out of the way. Throughout my years in the fire service, I have always been taught this principle and have been warned of the consequences if the line were to pull away and strike a firefighter. For a while I thought it was one of those accidents that was always talked about but never really happened. Then it did happen. A firefighter in my department sustained a severe leg injury when the 5" line that he was connecting to the hydrant got snagged in the hose bed, pulling the line out of his

Figure 1–4. Sometimes a simple supply line is not sufficient.

hands and away from the hydrant. He was struck in the leg and required surgery and six months of convalescence to get back on the job. A second accident caused by the same type of scenario pulled the wrap off the hydrant, and a firefighter was struck in the head by the hydrant adapter on the end of the supply line. Fortunately it only contacted his helmet, which protected him from injury.

These accidents can happen. Do not depend on remembering the proper sequence for physically wrapping the hydrant with the supply line. There is a safer method. It involves using a three-piece strap system called the *hydrant strap*. The hydrant strap consists of a seatbelt that connects to the supply line and a 6' looped section of 1½" strap that also hooks to the supply hose. It is used to secure the supply line to the fire hydrant while allowing the firefighter to stand clear. Attached to the seatbelt is a tool pouch designed to hold all the tools needed for the hydrant operation. When a line is laid, the firefighter simply grabs the looped portion of the strap, pulls over to the fire hydrant, places the loop around the hydrant, stands back, and signals the engine to lay a line.

For best results, two 10' folds are placed at the end of the supply hose in a bundle with the adapter winding up in the middle. This allows for extra hose at the hydrant to ensure that the hose will reach. The seatbelt portion of the strap system is secured just behind the coupling strapping the bundle together. The looped portion of the strap system is passed through the fold at the end of the supply line and secured through itself, thus forming the loop that is used to anchor the hose to the fire hydrant. This system allows for several different types of hose laying evolutions.

We have already talked about laying the line from the hydrant with one firefighter. What if we wanted to lay a dry line from the hydrant and allow a second incoming unit to hook it up and charge it? Who is best suited to do this job? Most fire apparatus now have air packs in the jump seats, and the firefighters are packing up and sometimes going on air before they even arrive at the fire, thus allowing for an immediate attack. This being the case, it is not only safer but also faster for the driver to get out, loop the hydrant with the hydrant strap, and get back in the rig to lay the line. The driver will not be weighted down with an air pack and likely will not be wearing turnouts.

Another scenario where this strap works well is for the hose lay when a hydrant is not involved, such as a reverse or a rural application in which you might be stretching hose down a long road or driveway for an incoming water shuttle operation. The strap can be used to anchor the supply line to a fixed object or even another unit.

Figure 1–5 (a–f) shows how the hose strap is used by looping part of the strap system that is passed through the fold at the end of the supply line and secured through itself, which forms the loop used to anchor the hose to the fire hydrant.

Finally, with our fire trucks getting taller every year, it is becoming increasingly difficult to pull hose from the hose bed. The hydrant strap system allows for the loop of the strap to hang over the edge of the rear of the hose bed so that the firefighter can pull it from the street without having to climb onto the rig, again making for a safe operation.

Deploying a supply line

Now that our line is secured to the hydrant, let's lay a line. One thing I have noticed in my travels is that many departments tend to baby the process of laying large diameter hose, especially 5" hose. For some reason, they feel it is important to lay the hose out extremely slowly, probably because of the size. This is not necessary. Do not be afraid that you will damage the hose because this hose is fairly tough. There is nothing wrong with laying a supply line as fast as 15 mph. The hose actually tends to lay onto the ground softer when it is laid at a higher speed. The line stretching out from the hose bed will be tighter than if the hose and couplings fall straight down from the hose bed due to a slower speed. The key to fast deployment is having an obstruction-free hose bed. If your hose bed has a crossbar across the top, is it really needed? We do not ride tailboard anymore. If a grab handle is needed to pull you into the hose bed, look for an alternative place where a side handle can be mounted onto the rig. Let's clear up the hose bed.

We all know how a supply line can turn into a roadblock once it has been charged. Granted, large diameter hose can be driven over by other fire apparatus

Figure 1–5. How to use a hose strap

as a last resort as long as the units have the clearance, but it is better to avoid doing so. If the fire is on the opposite side of the street from the water source, lay the line as far down as you can on the same side as the water source before crossing and blocking the street. This will allow units arriving after the line is laid to get as close as possible to the fire, which will eliminate the need for firefighters to carry equipment long distances.

LDH puller strap

If a LDH line should become a roadblock, a simple method has been developed to move a charged LDH line by one or two firefighters. This method involves placing a canvas strap around the charged line and simply pulling the line to the desired area [fig. 1–6 (a–c)]. The strap consists of a canvas piece of material about ¾" to 1" wide and 8' to 10' long, sewn into a big loop.

When placing the strap around the charged line, straddle the line and bend down, keeping your back straight. Lift the line just enough to pass the strap under it. The line can be held in the raised position by resting it between the heels of your feet. The hose will be light enough to allow it to be lifted in the above-mentioned manner safely. Next pass the strap under the hose and pull it around the hose and through the other end of the loop until all the slack is gone and the strap is tight against the hose. The line can now be pulled. Although one firefighter can pull the line, it is best to use two whenever possible. A hydrant wrench or some type of bar should be placed in the loop on the free end of the strap to provide a better grip for pulling.

This method for pulling a charged LDH line is not a cure-all, but it has proven helpful in many situations and is safe. Tests have proven that the line can usually be pulled around 10' to the side with no problems. Testing has also demonstrated that it is more difficult to move the line in the area of a sharp bend. Give this method a try and see if it can be useful for you.

LDH hookup to the engine

Now let's discuss making the supply line hookup to the engine. Obviously the first step is to pull the amount of hose needed from the hose bed to make the connection to the intake of the pump. Too often I see the firefighter who standing at the

(a) (b) (c)

Figure 1–6. Step-by-step use of the hose puller strap

tailboard of the engine pull the hose straight to the ground, forming a big pile of spaghetti (fig. 1–7). This mess has to be straightened out before the line can be charged for obvious reasons. I have actually seen the hose get pushed underneath the fire truck, forming severe kinks and completely shutting off the water supply. Do not rely on your memory, especially in the heat of battle, to remind you to straighten out the kinks in a supply line after it has been connected. I recommend laying the hose out as free of kinks as possible when pulling it from the hose bed (fig. 1–8). It has to be done anyway, so why not do it right away? Simply grab the hose and walk straight back, pulling it towards the left or right depending on which side you will make the hookup, until the appropriate coupling hits the tailboard. Next, disconnect the hose and make the intake hookup. If you do forget to straighten out the supply hose, with this technique there are only two kinks that will have to be dealt with. Often the supply line will straighten itself out.

Marking the hose

Ideally supply hose, especially large diameter hose, should be in 50' sections. This makes it much easier to pull from the bed to make the hookup. Unfortunately,

Figure 1–7. The wrong way to pull supply line from a truck bed

Figure 1–8. The right way to pull supply line from a truck bed

most of the large diameter hose is in 100' sections, making it somewhat difficult to work with when the last coupling in the hose lay ends up being quite close to the tailboard. Many departments carry 25' and 50' *pony sections*, or custom lengths of hose also called *shorties*,[4] to complete the supply line hookup at the intake when this situation arises, rather than pulling an extra 100' of hose from the bed. There are two ways to deal with this problem. One is to mark each length of hose at 25' intervals to indicate how much of the last piece of hose is left in the hose bed. This will in turn indicate whether a 25' or 50' pony is needed, or if the remaining hose is sufficient in length and can be pulled for the hookup instead.

My preferred method is to only use the pony sections if less than 25' of the last section of hose is on the ground. It is easier and less time-consuming to pull the remaining hose from the bed, making only one connection instead of pulling an extra piece of hose from a compartment and making two connections.

When making the hookup to the intake valve, make sure that you have a wide bend in the hose to eliminate any kinks that cannot be straightened out [fig. 1–9 (a–b)].

(a) (b)

Figure 1–9. Make a sweeping bend into the intake to avoid these kinds of kinks in the hose.

Charging the line

When large diameter hose is charged from a hydrant, the water moving through the hose pushes air. It is very important to bleed the air out before it reaches the pump (fig. 1–10). Failure to do so will allow the air to go through the pump and into existing handlines, temporarily creating pump *cavitation*, "a phenomenon in which rapid changes of pressure in a liquid lead to the formation of small vapor-filled cavities in places where the pressure is relatively low."[5] These areas of relatively lower pressure could cause handlines to go limp for several seconds.

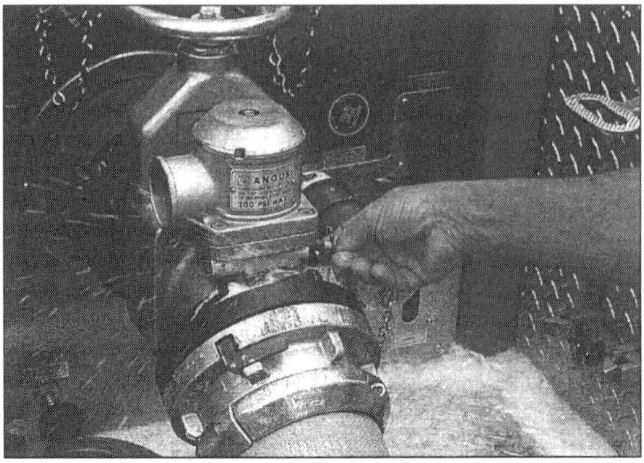

Figure 1–10. Bleeding air from a supply line avoids cavitation.

The longer the supply line, the more air could be present. It is a very good idea to make sure that the air bleeder on your intake valve is always open. Do not be fooled. Even though the valve is small, it can expel air fairly quickly. Instead of trying to remember to open the air bleeder valve in the heat of the battle, make it part of your morning routine to check that the valve is preset in the open mode.

Now that we have our complete supply line evolution laid out and connected into our apparatus, it is time to charge the line. Again, you do not have to baby the process and take forever to charge the line due to the size of the hose. Your goal should be to charge it as quickly as possible while not allowing the line to violently whip around. Charge it enough to loosen kinks, but not too quickly. You will not be able to totally eliminate all twisting of a supply line without expending too much time in doing so. Remember, this water is precious. We need to get it to the engine as safely and efficiently as possible.

Most departments do not use hose clamps in conjunction with large diameter hose. The firefighter at the hydrant is required to wait for a signal to charge the line. This can eat up valuable time, however, and could result in a pumper running out of water if the operation started from the booster tank. Precharging the supply line can reduce the time needed to get the supply line into service. As soon as the hydrant hookup is made, the firefighter can slowly start precharging the supply line. The line should be charged either to a point just before the line connects to the engine or to a point where the firefighter at the hydrant can no longer see the hose.

Replacing large diameter hose using the vacuum pickup method

Recently I taught a large diameter hose basics class to a group of small fire departments that had made a combined first-time purchase of 5" hose. Their plan was to get the new stuff loaded on their engines just prior to the day I was to teach so that all units would be ready. Two days before the scheduled class, however, they received notice of a big fire. Because the 5" hose was on the engines, it was used. It worked out well because the supply line that was needed was 1,300' long. These firefighters were able to move an impressive amount of water in a short time. The downside of the new and exciting experience was that they then had to pick up the 1,300' of 5" hose. They had not considered how different this process would be compared to picking up their 3" hose.

The first thing they realized was that the hose was heavier (110 lbs per 100' section of dry hose). After disconnecting each section of hose and draining the water, they thought that the hose could be loaded back on the engine just like the 3" hose. They were wrong. The firefighters soon found out that there was a little more to the process, since big hoses can retain a lot of air. In fact, they were not able to get all of the hose replaced in the hose bed. When I arrived, the firefighters looked frustrated. They were wondering if they had made the correct decision with their new hose purchase. I assured them that once they learned how to work with large diameter hose, it would be easier to pick up than the 3" hoseline.

The process of efficiently retrieving large diameter hose is based on a vacuum principle that allows water to be drained out of the hose while simultaneously preventing air from being pulled back in. This process draws the hose down completely flat and makes it easy to reload.

The first step in this process after the hydrant has been shut down is to disconnect both sides of the line, at the pumper and at the water source. Ideally you should not have to disconnect any of the couplings in between.

After both ends have been disconnected, seal off the end of the hose that will not be used to drain the water. This can be accomplished by either connecting a blind cap to the coupling or by recoupling the end of the hose back into the remaining hose in the hose bed. This should be done as soon as possible so air does not get drawn back into the hose.

After the hose has been sealed off, it can be drained by walking the water away from the sealed end. This can be accomplished using either a hose rolling tool or some type of bar. Either method will require a minimum of three firefighters to accomplish. Two firefighters will do the actual drain process, getting on each side of the tool. A third firefighter will stay behind, keeping the hose taut and thus allowing the water to drain freely. As the water drains out, a vacuum is created in the hose since the seal prevents air entering from behind. This in turn draws the hose down flat. As soon as the draining process reaches a point about 10' from the end of the line, stop, leaving the remaining water in the hose. The hose should be folded over a couple of times at that point so air cannot enter. The small amount of water left in the hose will help seal the hose at the fold.

At this point, if the cap has been used, the hose can now be loaded back onto the hose bed (fig. 1–11). Before removing the cap from the hose, place a fold several feet back from the end of the hose so air will not reenter the hose when the cap is removed (fig. 1–12). If the line has been sealed off by recoupling it back into the remaining hose in the hose bed, the above-mentioned step will not be needed.

When picking up the hose, the pumper can back into it, straddle it going forward, or run parallel to it going forward (figs. 1–13 and 1–14). Water can accumulate in the hose every so often, especially on longer lays. If this should happen, simply open the line at the next coupling to allow the water to drain out. Then reconnect the remaining hose and continue the process.

Figure 1–11. An end cap can be used to seal off the end of the hose.

Figure 1–12. An alternative to the end cap is to recouple the hose into the remaining hose in the hose bed.

Figure 1–13. The straddle method for loading the hose

Figure 1–14. The parallel method for loading the hose

The Hose Mule, a Lay-Flat Hose Retrieval System

The Hose Mule is a tool designed by a firefighter for firefighters. It is a great example of working to improve firefighter safety while developing equipment to do the heavy work.[6]

As mentioned previously in this chapter, larger diameter hose is capable of moving more water than the 2½" and 3" hose used in the past. And in one respect, it is easier to pick up because less hose has to be used. There is no getting around the weight of larger hoses, however, especially 5" hose. Do not forget that the hose must be picked up after the fire when firefighters are already exhausted, making the task even more difficult.

Clint Baker, fire engineer with the Temple, TX, Fire Department and CEO of Baker Fabrication, has developed a machine that has made replacement of large diameter hose in the hose bed of the truck easier and safer. It is called the *Hose Mule* (fig. 1–15).

Throughout Baker's career, he became familiar with both the advantages and disadvantage of large diameter hose. One of the main disadvantages of large diameter hose is the difficulty returning the hose into the hose bed. Baker's solution to this problem was the development of the Hose Mule.

The device

The Hose Mule is a motorized set of rollers designed to take the place of the firefighters who would otherwise have to manually pull the hose into the hose bed. Its operation simply allows the hose to pass through the set of powered wringers back into the hose bed. The Hose Mule is mounted on top of the apparatus either in the front or rear of the hose bed. The process involves three firefighters plus the driver, as well as one firefighter to perform the duties of the safety officer watching the operation. One firefighter operates the Hose Mule while two

Figure 1–15. The Hose Mule is mounted on top of the apparatus either in front of or to the rear of the hose bed.

Photo courtesy of Clint Baker.

firefighters pack the hose in the bed, which mainly involves creating the folds. It is helpful to have the hose predrained before the pickup operation starts, but it is not critical that all the water be evacuated.

The operational process for replacing the hose in the hose bed with the Hose Mule involves the apparatus moving forward alongside the hose at a speed of less than 5 mph (fig. 1–16). The crew in the hose bed should safely kneel in the bed while reloading the hose to maintain their stability and thus avoid a mishap. The Mule begins to pull the hose up to the hose bed while the apparatus is moving, and the remaining water can easily be drained out. When the coupling reaches the Hose Mule, the operator opens the wringers by lifting a handle mounted to the top portion of the machine. A platform mounted at the bottom of the machine lifts at the same time, elevating the couplings to the level of the rollers. At this point the firefighters simply pull the couplings past the roller and the rollers are lowered back against the hose, continuing the pulling operation. The average amount of hose that can be picked up per minute is between 50' and 80'.

The Hose Mule is compliant with *NFPA 1500: Standard on Fire Department Occupational Safety, Health, and Wellness Program*[7] for reloading large diameter hose onto a moving apparatus in regard to having firefighters on top of a moving apparatus for the purpose of replacing hose in the hose bed.

The Hose Mule can also pull in 1¾" and 2½" hose from a stationary position, which can then be reloaded directly into a hose bed. This can be of great benefit

Figure 1–16. The Hose Mule is a motorized set of rollers designed to replace the manual labor formerly necessary to pull hose into the hose bed.

Photo courtesy of Clint Baker.

to firefighters at a wildland fire when units have to quickly reload hose and go to the next job.

The Hose Mule comes in several models, allowing hose as large as 12" in diameter to be reloaded. It is powered by a $1/6$ horsepower gear motor and can either run directly from the apparatus battery or be hydraulically powered from a separate unit.

2

PUMPING FROM THE HYDRANT

Introduction

Pumping from the hydrant is a firefighting operation that has been conducted for many years. It involves placing the pumper at the hydrant, making hookups with short supply hose, and pumping through its discharge lines to another pumper or to water delivery appliances or handlines. Handlines are most commonly used when the hydrant is close to the fire.[1]

The main reason, at least hydraulically, to have the engine pump from the hydrant is to create enough pressure to move the water to the discharge delivery point. This operation is often used when the hose layout from the hydrant to the discharge point is very pressure restrictive. If this is the case it will be difficult to produce the required *hose stream*, or the "stream of water after it leaves the nozzle until it reaches its final destination, which is usually the seat of the fire" (fig. 2–1).[2] Pressure restrictions could include the diameter of the hose, the length of the lay, low hydrant residual pressure, and the volume of water, any of which could cause pressure loss. Before the introduction of large diameter hose, 2½" and 3" diameter hoses were normally used for supply line evolutions. Depending on the capacity of the hydrant system, pumpers on the hydrant were quite common even in low flow house fire scenarios. When the water requirements were high, it was also common to see multiple supply lines being pumped from the hydrant.

Fire departments quickly recognized the value of the low friction loss capabilities of large diameter hose. Hydrant pump operations faded away to the point where most hose evolutions did not include friction as a consideration. Basically, we got spoiled. On large flow fires, it is common to simply lay large diameter supply lines not only from the close hydrants initially used during the first alarm response, but even from the hydrants accessed hundreds of feet away.

As good as large diameter hose is for fire operations, it still has friction loss that must be overcome by a positive pressure source. Hydrant systems, even the

Figure 2–1. This 600' stretch of 5" supply line is powered from hydrant pressure. It performs more effectively if pumped from the hydrant.

so-called hot ones, are limited in their pressure capabilities. In fact, when a supply line supplied directly from a hydrant and pushed to the maximum output cannot provide any more water, it is usually not due to a lack of water. Instead it is because it ran out of system pressure. As mentioned in chapter 1, an important point to remember is that hydrant pressure starts at static pressure, which is the water in the system under normal conditions (not flowing appreciably). The static pressure is often thought of as the pressure that moves the water, but this is incorrect. Once the water starts flowing, this pressure becomes residual pressure and starts dropping. The more the water flows, the lower the pressure drops. This happens until you cannot access any more water because the pressure will not deliver it through the supply line.

The effects of static pressure and residual pressure can be illustrated with an example. Consider a hydrant with 100 psi static pressure. After flowing 1,000 gpm, the residual pressure is 40 psi. The friction loss in 5" hose is 5 psi per 100'. After leaving 20 psi for an intake pressure, the remaining available pressure will only move the volume of water 400' due to friction loss. The point of this story is that we can often move more water by creating additional pressure through the use of a pump at the hydrant.

Let's review some steps that can be taken to improve hydrant pumping capabilities. When making the hydrant hookup, it is important to access as much water as possible to support the units on the fireground. Often departments that use large diameter hose only make the connection with a single line (fig. 2–2). In fact, there are departments that only have the capability on their engines to use one line, leaving one of the main pump intake ports capped off. Why settle for

Figure 2–2. A hydrant hookup made with a single LDH line does not get the full capacity of the hydrant in many cases.

less when you have the capability to get more? You may not need the extra water, but it is better to be safe than sorry.

The most common hydrants in use today have three discharge ports. One port is large, usually 4", and the remaining ports are 2½". Many departments make multiple hookups with hoses matching the outlets. The large outlet gets the large line, and the 2½" outlet gets a 2½" or 3" line because their 2½" couplings match the outlet. At first this sounds sensible, but the true fact is that water follows the path of least resistance. With this combination of hose, the path of least resistance is the large outlet with the large hose. The smaller outlet with the small diameter hose will not get its full water supply.

A simple flow test that we conducted using a 500 gpm flow illustrates this concept. A flowmeter was used to measure the output from a 30' section of 3" diameter hoseline that was connected to the 2½" hydrant port. Flow was also measured from the large port, which was connected to a 50' length of 5" diameter hoseline (fig. 2–3). A water flow rate of 500 gpm was discharged from the engine, but only 60 gpm came through the 3" hoseline. The flow test demonstrated why it is wise to use large diameter hoses on every port, even the small ones.

I recommend using at least two lines for a hookup. In most cases this will get the hydrant capacity. However, if the hydrant is known to be exceptionally good, use a third line. You might be surprised what this could accomplish (fig. 2–4).

Believe it or not, the 2½" auxiliary suction under the pressure that would be provided in a relay can easily deliver between 500 and 800 gpm, verified by flow tests. We conducted a flow test with the Bayou Cane Fire Department in Louisiana to determine the flow rate capabilities of a 2½" intake. Using 4" large diameter hose, a relay was set up 500' away, and we were able to flow 800 gpm (fig. 2–5). So as an alternative to a large intake, if you need more water and do not have any other large diameter intake choices, a 2½" auxiliary intake delivering 800 gpm is far better than nothing.

Figure 2–3. This hookup is made with a 5" line and 2½" line, restricting the flow.

(a)

(b)

Figure 2–4. Multiple hydrant line hookups

(a)

(b)

(c)

Figure 2–5. A 4" relay allows 800 gpm to flow through a 2½" intake to provide more water.

Spotting the hydrant has always been one of the true tests of a pump operator's skills, with the objective of having a kink-free line after it is charged. Many years ago, soft suction lines were short for reasons that are no longer apparent. It was common to have lines as short as 12' to 15', and using them efficiently required skill and practice. I am all for having well-trained operators, but a short s does not help in this type of operation. The use of 25' and 50' sections makes better sense. The extra length does not affect the flow, but more importantly, it allows the unit placement to be simple. Basically all that is needed is to pull up within a reasonable distance to the hydrant, and the line will connect with no kink issues. If there are some initial kinks when the line is charged, they will most likely straighten out on their own or with a little assistance (fig. 2–6).

Figure 2–6. Deploying LDH safely and easily is done using the double donut roll. All the firefighter has to do is drag the roll instead of carrying it or unrolling it.

Big Stream Operations Require a Proactive Water Supply Approach

Fire departments are responsible for protecting their communities from the ravages of fires, both big and small. It would be ideal if all fires stayed in the small range, requiring only one or two handlines to extinguish. There are communities in which fires are often small, but even these communities have the potential for a large fire. It is with the small fires that we tend to have the most experience, both on the fireground and from our training. When it comes to water delivery on a small fire, operations are often basic. Fire attack is started from the onboard booster tank from the first-in engine, and if a sustained water supply is needed, a supply line is laid from the hydrant. Nearly all small fires can be handled with a flow rate of 500 gpm or less.[3]

So what happens when we do get the large fires? When it comes to water delivery, fire departments may struggle to develop the required flows. The problem does not stem from the required streams in terms of deploying the lines and appliances. Actually, that part of the operation is fairly simple. On large flow fires, *master streams* and large handlines are usually employed. A master stream is a water delivery appliance designed for large flows. It can be fixed, elevated, or portable. With large volume water delivery, the goal is to put the big ones into service, regardless of whether smoothbore tips or combination nozzles are used. This is only half the battle, however, and the required flow must be delivered from the water supply sources in order to support the big streams (fig. 2–7).

When you analyze water delivery operations from fire hydrants, the real problem is not usually a lack of water at the source. The water in a stored capacity in the water mains is almost always available, but the hydrant system pressure, or residual pressure, is often not sufficient to move the required flow over the required distance. In order to get the job done, multiple hydrants may be used, including hydrants on other grids further away in conjunction with specific pump operations. As long as the manpower, equipment, and determination to get the job done are all available, the required flow can usually be produced.

The fireground operation consists of several companies working in a coordinated effort under the direction of an incident commander or supervisor to execute a fire attack as safely and efficiently as possible. All phases of the operation must be conducted at the proper time to make the final outcome of the operation successful. This is where obtaining and successfully delivering the required flow from a water supply becomes crucial.

Often the request for more water is made when the existing water supply becomes depleted. Keep in mind that at a large-scale fire, there is a good

(a)

(b)

Figure 2–7. A secondary supply is needed because multiple large streams usually cannot be supplied from the initial fire hydrant.

possibility that multiple supply lines from the closest hydrants are already in place. Bringing in additional supply lines from new water sources will more than likely require long hose evolutions, preferably set up in a relay pump operation to achieve maximum flow. From the time the request comes in for additional water until the time the operation is set up and ready to deliver can easily take 10 to 15 minutes. If the current water supply is already depleted and frantic requests are coming in for more water, this will be too long to wait.

The best way to respond to a request for additional water is to be proactive. Do not wait until the operation needs more water; have it ready before the request is made (fig. 2–8). Setting up additional water supplies during the initial phases of a large flow operation will allow the overall water supply situation a better chance for success in fulfilling the required demands for the fire streams without interruption.

Here are two suggestions for proactively delivering additional water to the fireground. When command is established at a fire that has the potential to become a large flow operation, if possible one engine company from the first alarm assignment should be given the task of establishing a supplemental water supply. The engine should start its evolution from the fireground at a strategic location where the additional water may be needed. If the line is not ready to be connected to an engine, simply assign a firefighter to anchor the line. The engine should try to lay to a hydrant that is on a grid separate from the hydrants already being used. The engine should also set up at the hydrant for a relay pump operation. After making the hookups to the hydrant, the operator should charge the lines. Unless the water supply is already needed at the fire, however, the supply line should not be charged yet. If the line turns out not to be needed, it will be

Figure 2–8. Be proactive and lay a relay before you need it. Getting a relay ready for service takes longer than you can afford to wait when you are out of water.

Photo courtesy of Alan Simmons Productions.

much easier to pick up the dry hose than it will be to drain a charged line. The trade-off for this is very minor. Charging a supply line should only take a couple of minutes.

Another thing to consider in proactively seeking additional water is to make special requests in the dispatch assignment for units that will be wholly dedicated to bringing in additional water through supply line evolutions. Say for example a particular fire requires a three-alarm assignment to meet the challenge of the fire attack. Consider making a call for a fourth alarm, with the additional alarm being assigned to water supply. I understand that not all agencies have these kinds of resources. Having some type of plan for additional water supply, however, will improve the chances for a successful outcome on the fireground.

Figures 2–9 and 2–10 illustrate a first alarm and second alarm assignment. Note that the first alarm shows an engine that laid a proactive supply line set up for a relay. The receiving end of the line is not yet connected to an engine because at the time it was deployed, there was no need for the extra water. The second alarm shows the proactive relay connected to an engine to supplement the depleted water supply. There is also an additional relay set up with the second alarm assignment.

Supplemental Water Supply Operations

Obtaining a water supply

Obtaining a water supply on the fireground is one of those "gotta haves" for a successful outcome in combating a fire. For simple low flow fireground operations, this task is fairly easy. When fire hydrants are available, a single line from the hydrant will usually do the trick. But for more complex large flow fires that require large volumes of water to support master streams and big handlines, water supply operations can become quite difficult. Numerous factors can dictate how well a water supply operation works, including fire hydrant performance, fire hydrant availability, supply hose, available units, fireground topography, and water supply management, to name a few.[4]

We will discuss three different types of supplemental water supply operations for large flow water delivery involving multiple units:

1. Looped supply line operations
2. Manifold operations
3. Pumper supplying multiple large diameter supply lines

Maximizing the Water Supply on a Large Flow Incident with Supplemental Pump Operations

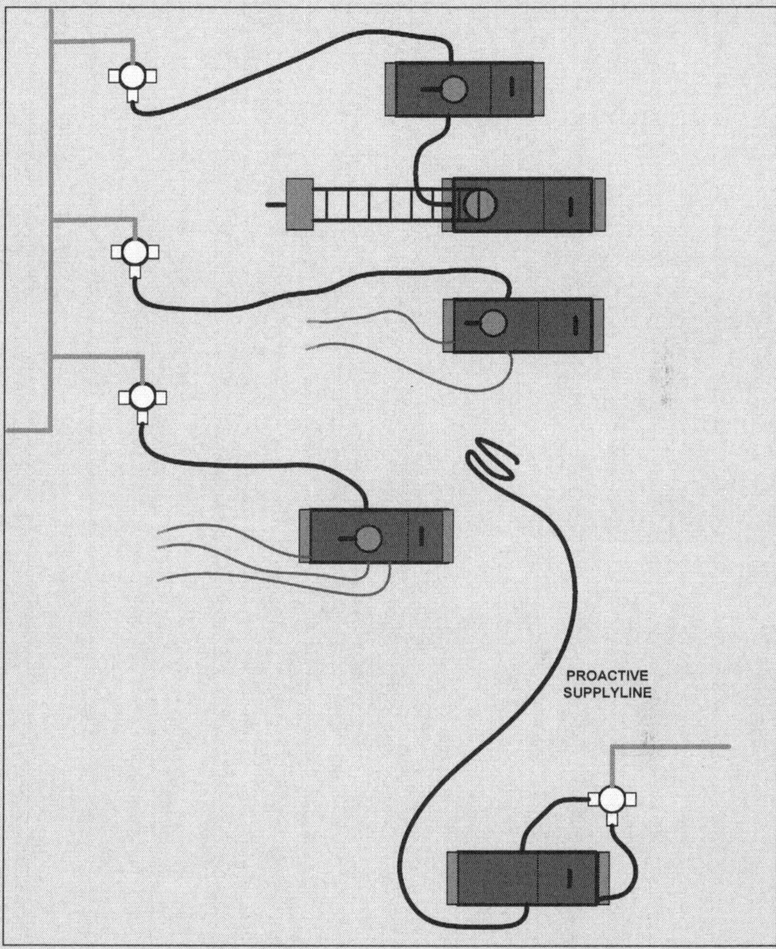

Figure 2–9. First alarm assignment. The relay is deployed but is not yet connected to an engine.

Maximizing the Water Supply on a Large Flow Incident with Supplemental Pump Operations

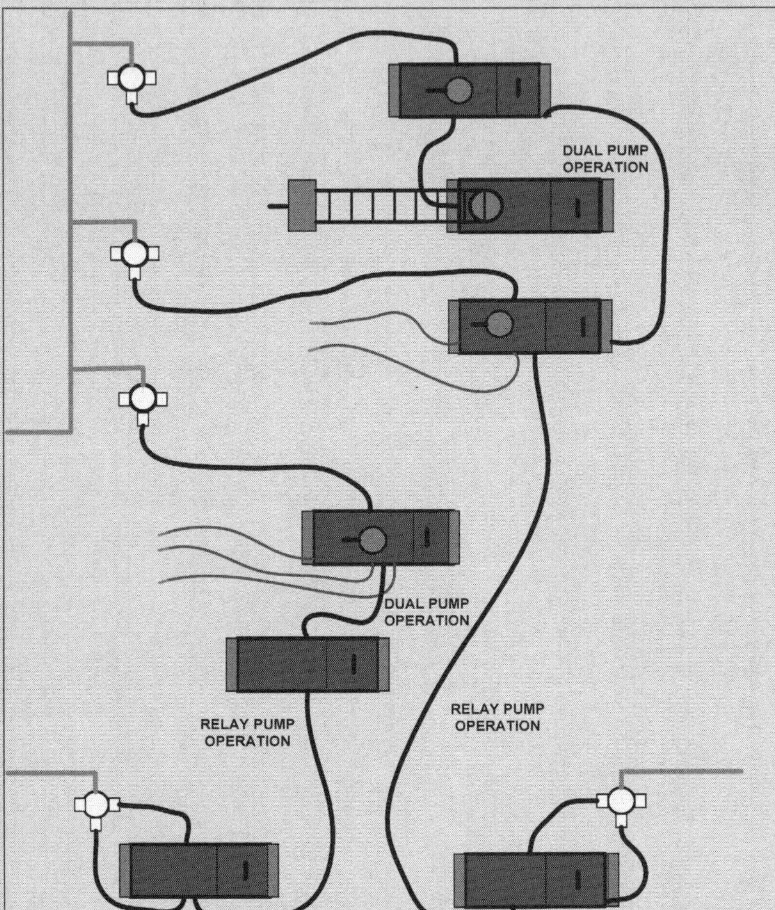

Figure 2–10. Second alarm assignment. A proactive relay is connected to an engine to supplement the depleted water supply.

A *supplemental water supply* is defined by the International Fire Service Training Association (IFSTA) as a method of operation for bringing in additional water to a pumper that is already connected to a water supply and needing more water.

Before discussing the three supplemental water supply operations, it may be helpful to review the basics concerning relay pumping.

Relay Pumping

Relay pumping is a pump operation that has been employed by the fire service for years. A relay pump operation is a supply line evolution that is required when there is not enough pressure from the water source to move the required volume of water to the fire through the supply hose.[5] This operation requires a pumper to be connected into the supply line evolution at the water source to boost the pressure in the supply line high enough to move the water to the pumper at the fire (fig. 2–11). This holds true for long- and short-distance operations.

Figure 2–11. A relay is set up through a hydrant assist valve hooked into an engine.

Photo courtesy of Paul Shapiro.

Relay pumping provides an excellent means for moving water from distant locations to the fire when there is an inadequate water supply or no water supply at all in the immediate fireground area.

A relay operation is probably the most complicated and dangerous pump operation that can be performed. It requires two or more pumpers to be connected into a supply line that has the possibility of sustaining high pressures and large volumes of water. Because of these high supply line demands, the potential for equipment failure due to hydraulic shock and pump cavitation is always present.

Through the years, procedures and equipment have been developed to make relay pump operations as safe and efficient as possible. Over time, however, things change. One equipment change that has taken place in supply line operations is the introduction of large diameter hose, and it has necessitated procedure changes in relay pumping.

Large diameter hose (4" and 5") has improved supply line water delivery capabilities significantly compared to smaller diameter hose (2½" and 3"). This significant improvement has actually eliminated the need to set up a relay in many instances. Along with the improved water delivery capability, however, large diameter hose also has a greater potential for developing hydraulic shock, or *water hammer*. Water hammer is defined as "a pressure surge or wave caused when a fluid... in motion is forced to stop or change direction suddenly."[6] The use of large diameter hose has thus led to the use of other new equipment and procedures to avoid water hammer and other potential problems.

This section will discuss updated information to assist in establishing safe and efficient large diameter hose relay pump operations.

Pump capacities as they relate to the placement in the relay

Through the years we learned that the largest capacity pumper must be located at the water source to guarantee maximum water delivery rates. This policy might be true when the water source is static, requiring a drafting operation. When the water source is pressurized through a hydrant system, however, this rule may not apply. To understand why this might be the case, it may be helpful to review the concept of net pump discharge pressure.

Net pump discharge pressure is the "actual amount of pressure being produced by the pump. When taking water from a hydrant, it is the difference between the intake pressure and the discharge pressure. When drafting, it is the sum of the intake pressure and the discharge pressure."[7] A pumper reaches its capacity rating when operating at 150 psi engine pressure when at draft. At draft, a pumper is creating its engine pressure while operating totally on net pump discharge pressure. A centrifugal fire pump, because of its design, increases its flow (measured in gallons per minute) as the net pump discharge pressure approaches 150 psi. Because of this characteristic, coupled with the reality of the high engine pressures needed in a relay operation, it is easy to understand why the largest capacity pumper should be at the water source during a drafting operation.

When a centrifugal pump receives its water from a positive pressure water source such as a fire hydrant, however, it takes advantage of the incoming pressure to actually increase the total pump capacity. How is this done? The incoming pressure adds to the net pump discharge pressure to produce the overall engine pressure. Because of this added help, the net pump discharge pressure (the amount produced by the engine alone) is lowered. Remember that the closer the net pump discharge pressure is to 150 psi, the higher the capacity of the pump will be. It is possible to double the pump capacity by hooking into a strong hydrant.

Given the type of flows that can be expected in a relay operation, coupled with the capabilities of the centrifugal pump and large diameter hose, there should not be any problem with placing a pumper with a flow rate as low as 1,000 gpm on the hydrant.

Use pressure dump valves to guard against excess pressure in a relay pump operation

Before the advent of large diameter hose in firefighting operations, unused discharge was employed to control excess pressure in a relay operation. With the introduction of large diameter hose, however, a new and preferred method for regulating the pressure has been developed—the *pressure dump valve*. The pressure dump valve is a pressure-regulating device designed to expel excess pressure from the water delivery system to the outside atmosphere. The pressure dump valve is connected directly to the large diameter hose system and can be field adjusted to the required pressure by the pump operator. Unlike the unused discharge, the intake dump valve requires no monitoring by the pump operator after it has been set to the proper pressure. The dump valve should be set as close to the residual pressure as possible to eliminate any problems from a potential pressure surge. Most dump valves are capable of discharging approximately 500 gpm. With the possibility of a surge larger than 500 gpm, a dump valve should be placed on the intake and discharge side of the pumper for added protection.

How to set pressure governors

Pumpers often come equipped with electronic pressure governors. Operators typically use them in the pressure mode, which works well for pressure surges and decreases, supply line changeovers, and the opening and closing of discharge lines. When two or more pumpers are operating in relay and the governors are all set in the pressure mode, however, there is a potential for problems. When the receiving pumper has a spike up or down in pressure, the other pumper reacts to it, causing a chain reaction of sorts that will continue back and forth between the units. The receiving pumper should be the only unit that is set in the pressure mode. All others should be in the engine speed (revolutions per minute or rpm) mode to help alleviate this problem.

The intake residual pressure for the receiving pumper or pumpers should never drop below the 20 psi safety margin. This margin allows the receiving pumper to keep extra water in the bank in case of a sudden pressure drop, thus reducing the possibility of cavitation to the pump.

General information regarding available flow rates in a relay operation

There are a number of guidelines that may be helpful concerning available flow rates in a relay operation:

1. The required flow rate (denoted in gallons per minute) must be available from the water source.
2. The following factors can have a direct effect on the amount of water that can be delivered in a relay:

    a. Pump size (if water source is static)
    b. Discharge plumbing size and design; plumbing designed for low friction loss at high flows will have the best results (fig. 2–12)
    c. Diameter of hose
    d. Operating pressure of hose (large diameter supply hose, 185 psi; large diameter attack hose, 275 psi)
    e. Length of lay
    f. Single or parallel lines
    g. Elevation gain or loss
    h. Number of pumpers involved in relay

Often fires requiring long-distance relays due to lack of hydrants in the immediate fireground area can be successfully extinguished with flow rates of 500 gpm or less. This being the case, as long as the water source is adequate and the length of the lay is reasonable, a large diameter hose relay operation in its simplest form will likely be adequate. In fact, if the water source is a strong hydrant, both in water volume and pressure, a simple forward lay without a relay might be feasible.

Structures that require large water flows for fire extinguishment hopefully also have been required by fire code to have an adequate hydrant system in the immediate area. This would eliminate the need for a relay operation. Since this is not always the case, however, there should be an alternative plan for getting additional water into the area. When organizing a water supply system, it is

Figure 2–12. A large diameter discharge flows maximum water most efficiently.

important to have the resources to allow the operation to grow if the fire problem escalates. Such resources should include the following:

1. Adequate water source or sources
2. Enough pumpers
3. Enough hose, which could include laying parallel lines the entire length of the hose lay and/or more than one evolution
4. Personnel trained to move water safely and efficiently

Hydrant relay valve

The hydrant relay valve (HRV) is a water delivery appliance that is connected to the hydrant and the beginning of the supply hose (fig. 2–13). It allows the initial supply line to be laid and charged, utilizing only hydrant pressure. If more water or pressure is needed, a second engine company can connect into the HRV at the hydrant and switch over a water control valve to the relay pump mode, pumping back into the supply line without shutting the hydrant down.

The hydrant relay valve was originally designed for areas lacking sufficient hydrant availability or placement. The valves allow the water supply to be maximized by setting up a relay. The trade-off for using an HRV is that it needs to be connected to the hydrant every time a pumper lays a line, whether it is used or

Figure 2–13. The hydrant relay valve (HRV) is preconnected to the end of the supply hose and is connected to the hydrant every time a supply line is laid, whether or not it is used in a relay.

not, because it is preconnected to the end of the supply line in the hose bed. These valves can weigh as much as 40 to 45 pounds. An alternative to using an HRV is to have the second pumper that will be doing the relay pumping reverse out from the receiving pumper at the fire and use another hydrant, setting up a relay. Either method requires two pumpers, but the second method will likely bring in more water.

The use of the HRV, as with any water flow appliance, must take into consideration associated friction loss, which will have a direct effect on the flow potential for the operation. There are different designs of HRVs, including the four-way hydrant valve manufactured by Humat. Some have more pronounced friction loss in the supply line mode (no pump at the hydrant, hydrant pressure only) (fig. 2–14), while others have greater friction loss in the relay pump mode (pump at the hydrant to set up in relay). Other HRVs exhibit friction loss in both modes. Based on flow tests, I consider the friction losses to be relatively insignificant in the HRVs. The small friction losses are well worth the trade-off when you consider what the HRVs allow us to do, namely hook up to an existing supply line without shutting down the hydrant (fig. 2–15).

Figure 2–14. The initial supply line is laid from the hydrant using the Humat valve.

Figure 2–15. Las Vegas Engine 4 uses the four-way Humat valve in a relay pump operation.

Short-distance relay pump operations

It has been an accepted practice to allow the shorter large diameter supply lines to be supplied solely from hydrant pressure. As mentioned in the previous chapter, firefighters compare laying a large diameter supply line to "taking the hydrant to the fire" or "laying an aboveground water main." In other words, there is an expectation that large diameter hose laid from a hydrant, especially on the shorter lays, will result in maximum flow (fig. 2–16). This type of thinking, however, can get us in trouble. Large diameter hose is great, but it still has restrictions concerning its use. Large diameter hose must be used effectively in conjunction with everything else on the fireground in order to reap its benefits.

In certain situations, a short relay pump operation could be an alternative to the supply line directly from hydrant pressure (fig. 2–17). You may be wondering why this relay pump operation should be set up when additional supply lines from other engines could be laid instead, allowing multiple pumpers to help deliver the water versus one pumper. The answer is that there are a number of situations that might require a single engine company to produce maximum flows. It may be that the area the water needs to be delivered from is only big enough for a single engine company to work from, such as a cul-de-sac, a parking lot full of cars, or an alley, to name a few constricted locations. Furthermore, what if additional engine companies are not responding or are extremely delayed? In such

Figure 2–16. Sparks Ladder 31 is in the background sitting on a hydrant at a distance of about 500' in a short relay pump operation.

Figure 2–17. This relay pump operation uses dual 5" supply lines.

cases, the "little engine company that could" will have to pump its little heart out. This is where short relay pump operations can be of great benefit.

Developing the proper engine pressure on the pumper at the water source

When forming a relay, it is nearly impossible to know ahead of time exactly how much water will be needed. It is also difficult to know how much hose has been deployed to make up the supply line, especially on the longer evolutions. This lack of information makes it difficult for the pump operator at the water source to determine the correct engine pressure.

To make it easier for the pump operator to establish the correct engine pressure, a *target engine pressure* should be used. The target engine pressure is a designated pressure that the pump operator automatically throttles up to in order to get the initial water to the receiving pumper at the fire. Often a target engine pressure of 100 psi is used. After water is flowing, an evaluation can be made on the existing water situation. Adjustments in engine pressures can then be made accordingly.

Relay pump operation procedures

As mentioned previously, a relay pump operation is probably the most complicated and potentially dangerous pump operation that can be performed. It is vital that pump operators be well trained in every aspect of the operation to ensure that the required water is delivered to the fireground safely and efficiently. Pump operators must understand why the pressure dump valve is needed and know how to properly adjust it after water is flowing. They should also have a good understanding of the principles of the movement of water in a relay operation in relation to all other equipment involved with the relay, such as the discharge being used, the type of hose, the capabilities of the pumpers, etc.

Note: Communication is extremely important between the pump operators involved with the operation. If manpower permits, designate a water supply officer to oversee the entire relay operation. It is also important to keep the fireground officer apprised concerning the status of the operation. This officer will need to know if the required amount of water can be obtained, and if not, whether there are any alternatives to consider. The fireground officer also should be given a basic summary of how the operation will be set up.

It should be noted that if the units involved with the relay have electronic governors, only the receiving pumper should be kept in the pressure (psi) mode. All others should be in the engine speed (rpm) mode. This will eliminate the back-and-forth engine speed changes that will occur with the water being opened and closed at the fire from the receiving pumper.

Let's take a step-by-step look at how a relay operation should be conducted:

1. After all the lines have been laid and connected, the pump operator on the receiving pumper should advise the pump operator at the water source that they are ready for water.
2. The pump operator at the water source opens the discharge valve slowly, allowing the relay supply line to fill completely under low pressure.
3. The pump operator on the receiving pumper should advise the pump operator on the water source when the water is received.
4. The pump operator at the water source should now throttle up to the target engine pressure (100 psi).
5. After the receiving pumper receives the initial volume of water at 100 psi and gets the attack lines into service, an evaluation should be made to determine the remaining available water status. The following questions should be asked:

    a. Is more water needed?
    b. Is the water I am receiving sufficient?
    c. Am I receiving too much water?

    Besides meeting the water requirements for the evolution, the final determination for the engine pressure should be made by noting the residual pressure on the receiving pumper in relation to the lowest setting that the intake pressure dump valve can be set at. The goal is to set the intake pressure dump valve as close as possible to the residual pressure to guard against pressure surges.

Example. The receiving pumper in a relay is equipped with a pressure dump valve with a minimum pressure setting of 50 psi. After this pumper is flowing its initial lines, the residual pressure has dropped down to 80 psi. The pumper needs to place two more 2½" lines into service, which will require 500 gpm more flow. This additional flow will drop the residual pressure well below the minimum of 50 psi. In this situation the pumper at the water source needs to increase the engine pressure to provide the additional flow while maintaining a residual pressure of 50 psi at the receiving pumper.

Now let's turn the situation around. What happens when the receiving pumper's residual pressure is 120 psi after the water requirements are met? Again, the receiving pumper is equipped with a pressure dump valve with a minimum setting of 50 psi. For the sake of efficiency, the engine pressure on the pumper at the water source should be lowered to 50 psi, while maintaining the required flow.

6. After the required amount of water is flowing and the engine pressures are set, the intake and discharge pressure dump valves should be adjusted as close as possible to the pressure at the valve to guard against any pressure surges.
7. When shutting down the operation, the receiving pumper should shut down first.

There are many different variables that may influence fireground operations that can have a direct effect on the type of water supply that will have to be established. A good preplan of your area, together with some excellent training, will get your department ready to tackle any water supply challenges that may arise.

Relay pumping can be a good secondary evolution that will ensure adequate water to a fire suppression operation (fig. 2–18).

Looped Supply Line Operations

The first supplemental water supply operation we will discuss is what I like to call a *looped supply line operation*. A looped supply line operation is a combination of supply line hose evolutions set up to work as one. Large flow operations

Figure 2–18. This 500' relay is utilizing 5" hose to support multiple master streams.

usually require multiple supply lines from various locations, all working toward the goal of providing an overall water supply for the incident. These operations can be quite complex and taxing because the overall goal is to provide the required flow to multiple units without interruption. Even given this goal it can be extremely difficult to implement a water supply operation and accomplish it without issues.[8]

The looped supply line operation is designed to deliver a supplemental water supply to as many pumpers as possible based on the placement of the apparatus at the fire (fig. 2–19). Think of this as a fire hydrant water main system. A single supply line connected to a pumper is like a fire hydrant connected to a dead-end main, only getting water from one supply, the main that is feeding it. The hydrant that is connected in the looped grid system is actually receiving additional water

Figure 2–19. This looped supply line operation by the Los Angeles City FD is necessary in a large flow operation.

from another point in the grid system, thus the term "supplemental." If it works with underground water mains, it should work above the ground with supply hose as well.

Dual pumping

Dual pumping is a pump operation that is a key factor in building a looped supply line operation. This operation is based on a water-sharing evolution designed to share water between pumpers (fig. 2–20).[9]

Water sharing has been conducted over the years via one pumper discharge pumping into another. This technique works and can usually move slightly more water than a dual pumping operation. So why should we use a dual pumping operation? The answer lies in the ability of a dual pumping operation to accommodate different directional water needs. When you discharge the pump through a hoseline, the water only goes in one direction, from discharge to intake. This can hinder a true looped operation. Dual pumping, believe it or not, is an operation that connects two or more pumpers together through the unused *intakes* of the pumps involved to share the water (intake to intake). Because there is no actual discharge pumping in the dual pump operation, the water can flow in either direction. The water will flow in the direction of the pumper that has the lowest residual pressure.

Figure 2–20. Dual pumping is designed to share water between pumpers.

Summary of Dual Pump Operations

1. Dual pumping moves water from one pumper to another with supply hose connected intake to intake.
2. The water is moved by the residual pressure of the first pump.
3. Dual pumping works well for low flow operations such as handlines when the water source to the initial pumper is coming directly from a hydrant with hydrant pressure only.
4. If a dual pump operation is going to be used to share big water, the water source to the initial pump needs to be boosted from the hydrant with a relay pump operation.
5. The placement of a dual pump operation is based on the flow requirements and the available residual pressure.
6. To avoid cavitation, the intake valves should be opened slowly until the line is charged.
7. Since dual pumping shares residual pressure, pump operators need to monitor the pump discharge pressures in case there is a drop in residual pressure.

Source: Paul Shapiro, "Supplemental Water Supply Operations," *Fire Rescue* 10, no. 11 (November 1, 2015).

For example, consider a pumper that lays in a 5" supply line and goes to work. The supply line has the capability of moving 1,500 gpm, and the pumper that laid the line is going to flow 500 gpm. After flowing 500 gpm, the intake gauge of the pumper reads 60 psi. The 60 psi is the residual pressure of the water supply left over after the 500 gpm is flowing. The 60 psi represents water under pressure from the hydrant that is available within the supply line and the intake side of the pump. Now let's say this first-in pumper wants to share some of its water with another pumper. A line is connected from its unused intake to the intake of the pumper it is going share water with. When the command is given to share the water, the first pumper opens its intake valve, which is connected to the dual pump line, and the water will start to move through the supply hose to the next pumper under the 60 psi residual pressure. This is somewhat similar to being connected to a hydrant, since the 60 psi residual pressure is actually coming from the hydrant. The valve is located before the impeller of the first pumper's pump, and therefore does not have anything to do with the pumping capabilities of the first pumper. When the water is received by the second pumper, the pump operator simply opens up the intake valve and gets the shared water from the first pumper.

One of the unique things about this operation is that the pumper that is sharing the water with the next pumper can never be robbed of water it is discharging. Due to the pump's design, all the water that comes into the first pumper will remain available for that unit when it is needed. Let's go back to the initial part of this example. After the first pumper is flowing 500 gpm and sharing water through a dual pump operation to the next pumper, the first pumper has to increase its flow by another 500 gpm. The only thing the pump operator will have to do is open the proper discharge lines and throttle up to the proper pressure to flow that extra 500 gpm. The water will basically be taken back from the dual pump operation to the second pumper. Again, the initial pumper receiving water from the water supply will always have priority.

There are a couple of factors that must be monitored during a dual pump operation by both pump operators. Once the initial pumper shares the remaining water supply, it will lose some residual pressure because the residual pressure is being used to move the water to the next pumper. Consequently, the pump operator on the first pumper might have to apply more throttle to increase the pressure back to whatever is required. If the pump is equipped with an electronic governor, it will automatically adjust the engine speed (rpms) to the new residual pressure. Again, the first pumper is not losing any water, only residual pressure. If the first pumper needs to take back some water, the pump operator should communicate with the pumper that is receiving the extra water that they will be taking water back, which will in turn drop the intake pressure of the receiving pumper. The pump operator on the second pumper should try to maintain the engine pressure that was required for its flow and may have to throttle up because of a drop in the intake pressure. Figure 2–21 illustrates a basic dual pump operation.

Dual pumping may be a new concept to some firefighters, so I recommend that you go out and train with it before integrating it into a looped supply line multicompany operation. Get comfortable with its use and see what it can do for your fireground operations.

Figure 2–22 is the beginning of a multicompany large flow operation and shows three pumpers connected to their initial water sources.

This is basically the first alarm assignment. These units are in the big water mode, so they are starting to tax the main that is connected to the three hydrants. It is realistic to expect that the first-in companies will grab the closest hydrants. There is the possibility that some or even all of the hydrants are on the same main. Now is when these first-in units are likely to start running low on water. Running low on water in and of itself is not a risk, as long as the engines do not try to deliver more water than their current discharge evolutions. Since the units in the first

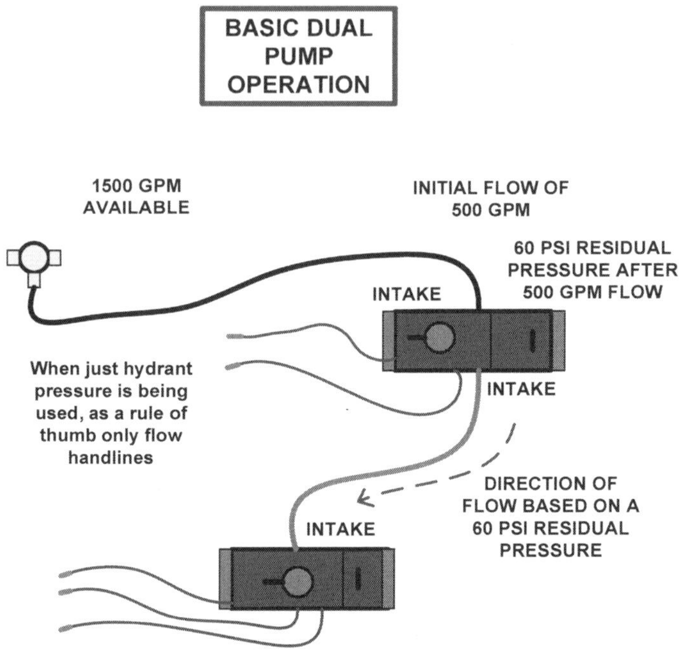

Figure 2–21. For big flows, dual pumping needs to be powered from a relay pump operation for higher pressures.

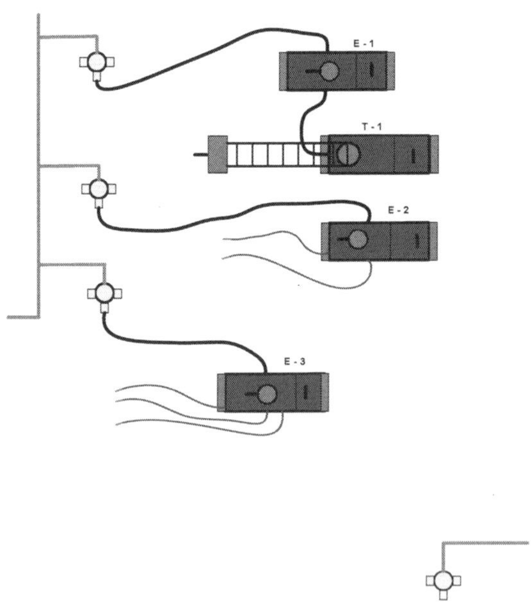

Figure 2–22. Three pumpers connected to their water sources

alarm assignment are usually in a good location to deploy multiple streams, however, there is a good chance the first alarm assignment will need more water.

The next illustration (fig. 2–23) shows supplemental pumping using the looped supply line operation, which involves relay pump operations and looping or connecting the receiving pumpers together to share a common water supply. Remember these units already have water coming into them from their initial supply lines. They just need to freshen up their water supplies. In some cases, they will not need much more water, while in others they may need to double their current flow. As mentioned previously, when units are connected in a dual pumping operation, the water will automatically flow in the direction of the pump that has the lowest residual pressure.

I am a big fan of preparing for worst-case scenarios. In regard to this operation, it is best if the pumpers are equipped with at least three large intakes—the two steamers coming off the pump and either a front or rear suction (fig. 2–24).

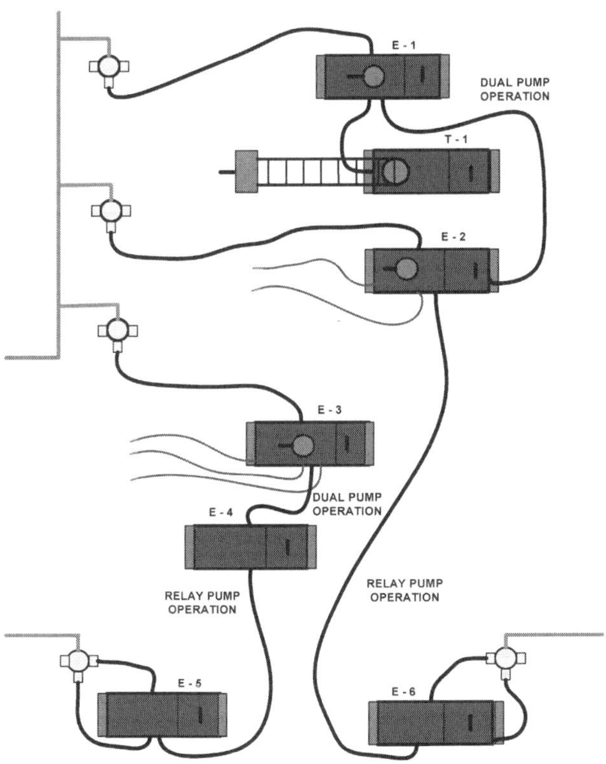

Figure 2–23. A common water supply is shared through looping a supply line connecting relay and dual pump operations.

Figure 2–24. Two large diameter manifolds can be put together to create three large diameter discharges.

There are manufacturers that make valves that can be connected to the main steamer that will allow two supply lines to be connected. The illustration shows two dual pump operations using pumpers with two intakes and pumpers with three intakes. Both will work, but the pumpers with three intakes allow for more options.

Manifold Operations

The second method for supplementing a water supply at a large flow operation is to use large diameter hose manifolds in conjunction with a relay pump operation. Remember in a supplemental operation we are giving additional water to units that already have their own supply lines but are running low on water and need more. There are two types of large diameter hose manifolds that are available (fig. 2–25). One has a large diameter hose inlet and outlet as well as multiple 2½" outlets. This style is designed more for handline operations supported by large diameter hoselines. In our situation, where we are attempting to move large volumes of water only, the 2½" discharge ports on the manifold will be very restricting and therefore should not be used for this type of operation (fig. 2–26).[10]

The other manifold that I recommend has a large diameter hose inlet and two large diameter hose outlets. The body of these manifolds is large and has unrestricted waterways, and it is very efficient for moving big water. These manifolds work best in conjunction with a relay pump operation. The manifold should be placed on the fireground close to the units that it will be supplying and can be fed by a single large diameter hose relay line.

Figure 2–25. Large diameter hose can be connected to the 2½" ports on the manifold; however, the 2½" waterways are very restrictive.

Figure 2–26. A large diameter hose manifold with large diameter inlet and outlets is very efficient.

Hydrant Performance

Reasons for poor hydrant performance include the following (fig. 2–27):

- Lack of hydrants
- Low-volume and/or low-pressure hydrants
- Small mains
- Dead-end mains
- Use of multiple hydrants from one main

(a)

(b)

Figure 2–27. Reasons for poor hydrant performance

Photo (a) courtesy of Alan Simmons Productions.

Although this is an efficient operation, it is important to remember that the water is being discharged through the entire hose evolution, and therefore will only flow the water in one direction.

A Pumper Supplying Multiple Large Diameter Supply Lines

The third method for supplying a supplemental water supply to units on the fireground is by simply pumping multiple large diameter hose discharge lines to the individual units. This can be done either from the hydrant or at the end of a relay on the fireground. Most units only have one large diameter discharge, but the 2½" discharge can be adapted to a large diameter hoseline that can move 1,000 gpm fairly efficiently.[11]

Flow Tests

Mustang Fire Department flow tests

Flow tests were conducted with the Mustang FD and Oklahoma City FD (fig. 2–28).

MUSTANG FIRE DEARTMENT LARGE DIAMETER HOSE SUPPLY LINE FLOW TEST 8/4/17

The purpose of these flow tests is to show the options available for the Mustang Fire Department from municipal hydrants using 5" hose. It should be noted that one test was conducted with Oklahoma City Fire Department engine 8 using 4" LDH. The static pressure for the hydrant being used is 60 psi.

TEST One				
200 feet of 5" hose from the large port on the hydrant to the side 5" intake valve.				
Deck gun	2" tip	70 psi NP	997 GPM	
Portable master stream	1-3/8" tip	48 psi NP	389 GPM	
Total flow 1386 GPM	Residual pressure for total flow 10 psi.			
TEST Two				
200 feet of 5" hose from the 2-1/2" port on the hydrant to the side 5" intake valve.				
Deck gun	2" tip	70 psi NP	997 GPM	
Residual pressure 10 psi				
Test Three				
Two lines 200 feet of 5"hose from the large port and 2-1/2" port on the hydrant into the side 5" intakes. (dual hose lay)				
Deck gun	2" tip	70 psi NP	997 GPM	40 psi residual pressure
Two Portable master streams	1-3/8" tip	48 psi NP each	389 GPM each	
Total flow 1775 GPM	Residual pressure for total flow 10 psi			
TEST Four				
Oklahoma City Fire Department 200 feet of 4" hose from the large port on the hydrant to the side 5" intake valve.				
Deck gun	2" tip	58 psi NP	905 GPM	10 psi residual pressure

Figure 2–28. Mustang (OK) FD and OKC FD flow tests

Woodinville Fire Department flow tests

Flow tests were conducted with the Woodinville FD (figs. 2–29 and 2–30).

Figure 2–29. Woodinville FD flow tests

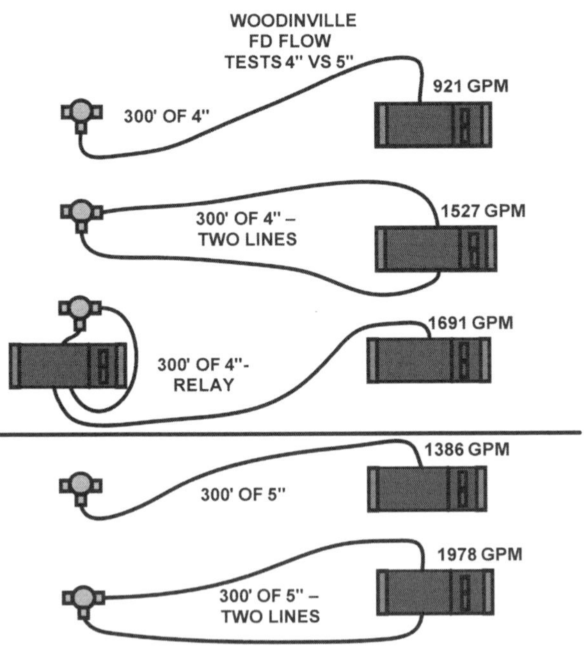

Figure 2–30. Woodinville FD flow test diagram

Fort Worth Fire Department flow tests

Flow tests were conducted with the Fort Worth Fire Department (figs. 2–31 and 2–32).

Figure 2–31. Ft. Worth FD flow tests

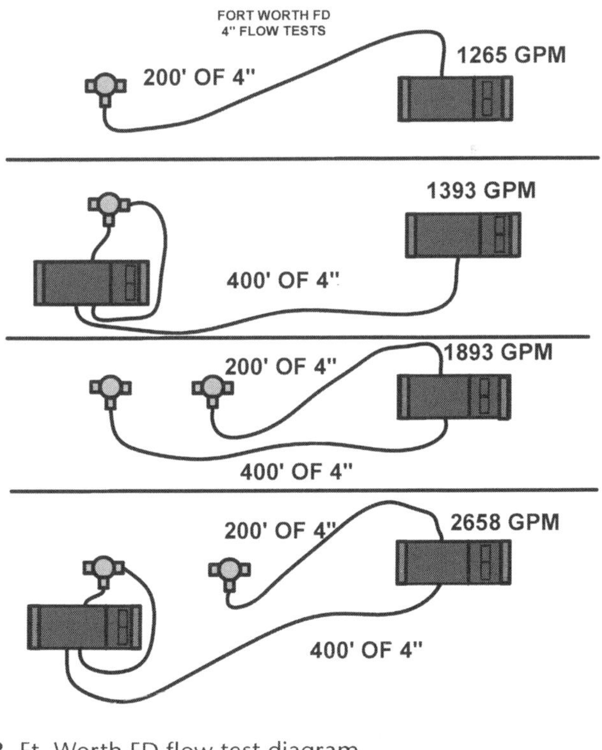

Figure 2–32. Ft. Worth FD flow test diagram

Las Vegas Fire Department flow tests

Flow tests were conducted with the Las Vegas Fire Department (figs. 2–33 and 2–34).

Figure 2–33. Las Vegas FD flow tests

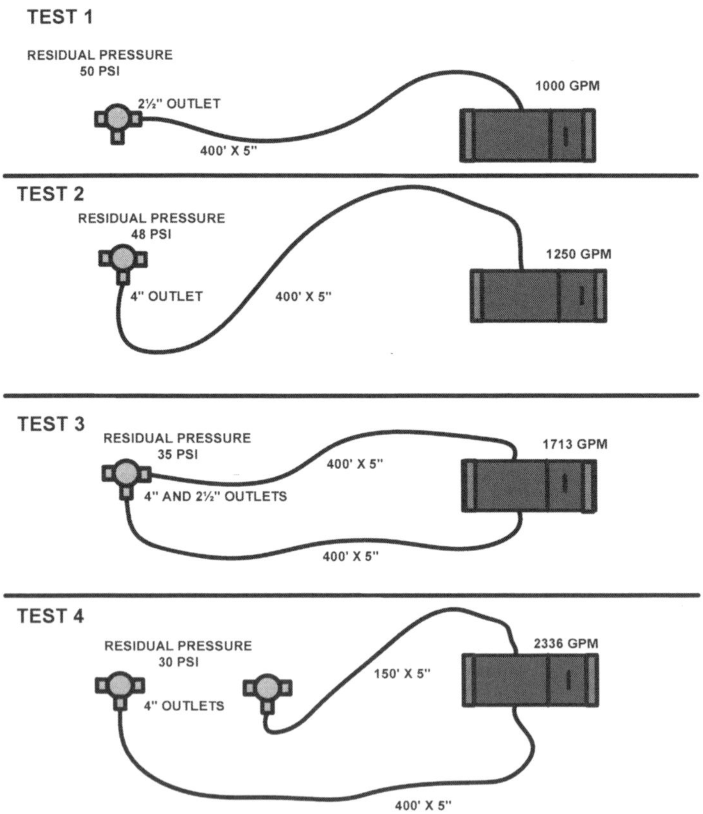

Figure 2–34. Las Vegas FD flow test diagram

Santiago (Chile) Fire Department flow test

Flow tests were conducted with the Santiago Fire Department (fig. 2–35 and fig. 2–36).

Figure 2–35. Santiago FD

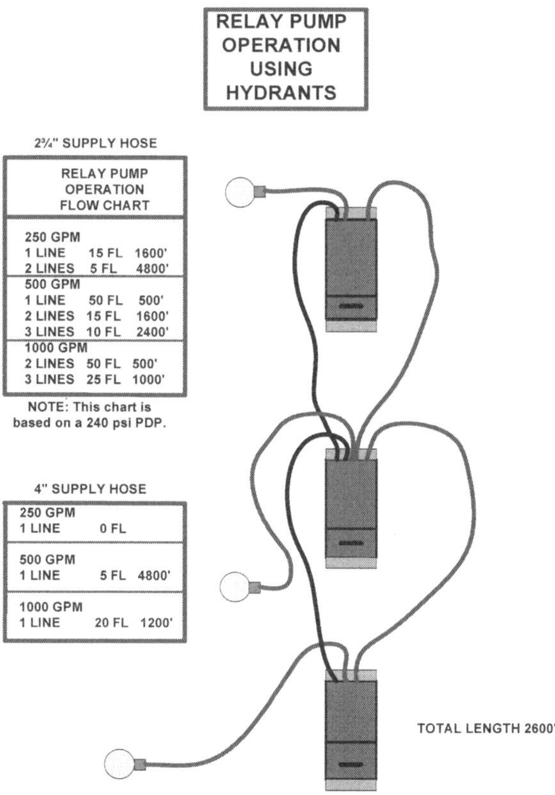

Figure 2–36. Santiago FD flow test diagram

3

STANDARD UNITS USED FOR BIG FLOW OPERATIONS—YOU ALREADY OWN THEM

Introduction

When we respond to a fire, we are expected to extinguish it in a quick and efficient manner, and the fire service is trained to do this well. Nothing is perfect, however, and every once in a while we experience unexpected problems that challenge us. One such challenge occurs with large flow fires. Such fires require not only that we ensure sufficient water from the water supply to overwhelm the heat energy generated from the fire, but also that we deliver or discharge the water onto the fire to get the knockdown as quickly as possible. This is where fires can go bad.[1]

Most of us are familiar with the big "super pumpers" that are used in refineries across the nation and even all over the globe. These heavy-duty pumper trucks are designed to flow large volumes of water, in some cases with flow rates in excess of 5,000 gpm, and are used to protect property (in this case, oil) worth billions of dollars. The refining facilities are set up with water delivery capable of supporting the flows from the Super Pumpers. In fact, one of the leaders in industrial pumpers makes a unit called the Inundator that is capable of flowing 10,000 gpm when utilizing a pressurized water delivery system.[2]

There are also units with pumps rated up to 3,000 gpm that can produce flows approaching 5,000 gpm when used in the appropriate pressurized water delivery evolutions. Remember, pumps are rated at draft at 150 psi net pressure. When a pressurized incoming water supply is used, it subtracts its incoming pressure from the overall pressure, thus allowing the pump to produce higher-than-rated flows. For example, consider a pump with a pump discharge pressure of 200 psi and an incoming pressure of 60 psi. By subtracting the 60 psi from the 200 psi, the net pressure will be 140 psi. This being said, the pump has not maxed out its flow capacity capabilities because it has not reached the

150 net psi mark. It is possible to double the rated capacity of a pump under the right circumstances.

Clearly water flow rates of this magnitude could be beneficial at a number of different fire scenes, including fully involved warehouse fires or construction fires. This level of water delivery capability generally will be useful at any fire that requires flow rates of thousands of gallons per minute of water to control and extinguish the fire as quickly as possible.

Currently when we are facing a large fire, we set up our standard master streams. If we max them out, which we usually do not, we are basically flowing 1,000 gpm. You may be wondering if it would be simpler to achieve the needed flow volumes with multiple streams instead of one larger stream. If we are flowing three 1,000 gpm streams, then we are flowing 3,000 gpm. So why do we need a single unit that can flow that much water? While it is true that both operations can achieve water flow rates of 3,000 gpm, the areas covered by multiple pumpers are usually more widespread. With multiple pumpers, we may be flowing three 1,000 gpm streams, but each one may only cover a portion of the fire. This often is not enough water in one location to knock the fire down.

The big stream concept of water delivery is not new. If you research the municipal big stream concept, you will discover that some large US cities built apparatus during the 1960s that could flow 5,000 gpm or more and were highly successfully in controlling large fires.[3] So why are we not using super pumper–style engines like this for municipal fire departments? One consideration is cost. Super pumpers are specialty units that are only used on big fires. These units are extremely expensive, and it may be years before they are needed for a big fire. Do we have a water delivery system that will flow 10,000 gpm to a single pumper? Yes, we have the technology. Do we know how to move that kind of water to a specific location? The answer to this is less certain.

Here are a few additional reasons why super pumpers may not be widely used:

- *Change of administration and priorities.* New leadership may change the focus of budget allocations.
- *Lack of personnel and funds to take on additional tasks.* Many fire departments already are tasked with EMS, Hazmat, Swift Water, and other responses and may face other community challenges as well.
- *Failure to teach the importance of water flow (gpm) vs. heat (Btu).* The volume of water applied must meet or overwhelm the heat energy of the fire. The practical ramifications of this concept are not always well understood.
- *Ignorance.* There may be a mindset that the status quo is sufficient. "We all have 1,000 gpm deck guns. That should be enough. If we can't stop it with that, it's unstoppable."

There are a couple of alternatives to super pumpers that can be effective at a fraction of the cost. One alternative is to use a pumper that you already own that can still be used as an everyday pumper running out of its own station in its own district. The other is a quick attack unit, which is a pickup truck with a fixed master stream mounted to it that can be used as an everyday service truck to run the department's deliveries to the stations.

Economy Super Pumpers

First let's talk about the pumper. Based on the water delivery evolutions that will be described in this chapter, a 1,250 gpm or 1,500 gpm pump can provide 2,000 to 3,000 gpm with the correct equipment and supply line evolution. First you would need a master stream appliance that is designed with a minimum of a 4" waterway and a 4" discharge from the pump using a 4" valve (fig. 3–1).[4]

We conducted flow tests using the Stang Master Stream, which operates best at 100 psi nozzle pressure. The nozzle pressures and tip sizes for the smoothbore nozzles are given in table 3–1. These are the optimum pressures, but other nozzle pressures will also work, as will be noted throughout this section. The corresponding nozzle reactions are also given in table 3–1. *Nozzle reaction* is defined as "the

Figure 3–1. This 1985 American La France 1,250 gpm is capable of 3,000 gpm through its 4" Stang gun.

Table 3–1. Stang Master Stream flow test data

Flow Rate (gpm), Actual	Nozzle Pressure (psi)	Flow Rate (gpm), Rounded Off	Tip Size	Nozzle Reaction (lbs reactive force)
2,247	100	2,250	2¾"	1,187
2,674	100	2,675	3"	1,413
3,010	92	3,000	3¼"	1,658

Note: Stang recommends 100 psi nozzle pressure for these tips for maximum performance.

Top Tip: Super Pumpers

In my opinion, let the pros do the installation work for your Super Pumper project. Many apparatus refurbishing companies offer this service. They employ professional designers and fabricators who know how to build the unit to counteract the high nozzle reaction that will occur from the high-volume flows.

force exerted on a firefighter or other anchor by a stationary spraying nozzle supplied by a flexible hose. The reaction direction is opposite that of the jet."[5]

The water delivery operation needed to deliver the water to the super pumper requires multiple relays. Up to five large diameter hose inlets could be needed depending on the hydrant performance, the size of the hose, and the required flow. The intakes on the engines are used first since they are available and go directly to the pump. If the supply lines exceed the available intakes, a manifold system should be used to receive them. If this is the situation, one intake on the engine should remain available to connect to the discharge of the manifold. If the valve is placed on the ground, a 20' section of 5" hose connected from the appliance into the intake on the pump will be needed. It is important to lay the 20' section of 5" hose at a 90° angle from the pump intake because when the relay lines are initially charged, they tend to push the manifold forward. At a 90° angle, a kink in the hose could still be created, but it should be easy to straighten. If the line is laid straight out from the pump intake, the valve and kink could be pushed under the pumper. At that point it will most likely be stuck, and it will not be possible to straighten it out. This will severely affect the flow.

In our testing we found that using the type of manifold that Kochek provided on the supply side of the evolution made a difference (fig. 3–2). The Street Fighter manifold has a large diameter inlet and multiple outlets extending from either side. This design was efficient, and placing the manifold on the supply side of the evolution made a difference.

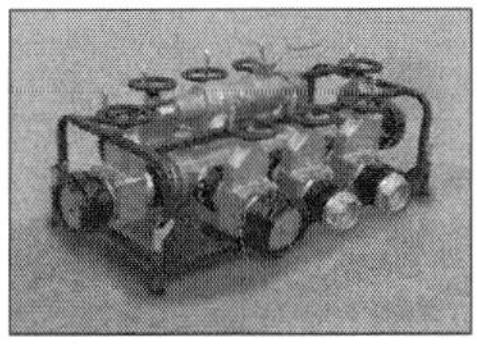

Figure 3–2. Kochek Street Fighter manifold, designed to fit the needs of our super pumper project.

The 4" gun can be used for standard flows as well as high flows (fig. 3–3). Unlike industrial pumpers, the engine of the apparatus does not have to be 600 horsepower (hp) because it is supported by the relay pumpers, which share the workload.

Setting up the economy super pumper

I believe there is a place for this type of unit within city departments, with some modifications in the unit and the water delivery operations required to support it. Let's take a look at how this can be done.[6]

First, I would make this unit capable of running as a regular engine company assigned to its own station and district, with the modifications previously mentioned to make it a high-flow pumper. The super pumper should have at least a 1,250 gpm to 1,500 gpm pump (fig. 3–4).

Both sizes of hose will work in a 3,000 gpm evolution, but with 4" hose, you basically have to work twice as hard. Two 4" lines equal approximately one 5" line hydraulically. Other than that, the unit can be designed and equipped as a standard engine company to run the standard calls.

Figure 3–3. The stack tips on this 4" gun start at 1⅜" and go up to 2¾".

Figure 3–4. This 4" gun is flowing 2,500 gpm from a 1,250 gpm pump supported by one 6" relay and its own hydrant.

If there is a big fire that requires this unit to go into high-volume mode, it can be called specifically for the incident if it is not already assigned. As mentioned previously, the unit would need multiple supply lines to develop a 3,000 gpm stream. If 5" hose is used, I would recommend a minimum of three supply line evolutions from three separate hydrants set up in a relay pump operation no matter how close or far away the hydrant may be from the super pumper. We attempted four flow tests using two hydrants, 5" hose, and two pumpers. We were not able to get enough water from the hydrants, and we ran out of throttle. The third pumper with a third hydrant made a difference. The extra pump reduced the overall engine speed for the other two engines, making the operation more efficient and achievable (fig. 3–5).

If the hose being used is 4", three hydrants would likely be needed. Two of them would be set up in a dual 4" relay line and one in a single relay operation. If single supply lines are used instead of duals, plan on using up to five hydrants because of the extra friction loss in the supply hose. The super pumper operation in itself would probably qualify for an extra alarm just to set it up, so it might be wise to include a special callout operating procedure if the super pumper is needed. The pumpers hooking up to the hydrants for the relay operation should hook up

Figure 3–5. This was originally a two-pumper operation set up to move the required water, but the engine speeds were too high. Addition of a third engine in tandem relieved some of the workload of the first two pumpers involved in the operation.

a minimum of two soft suctions to get maximum flow from the hydrant, and both of them should be large diameter hose. The 2½" port on the hydrant can be adapted to large diameter hose to connect to the intake of the engine. If you think the hydrant is strong enough and the unit is equipped for three lines, I would recommend making a three-line hookup to the hydrant for maximum flow (fig. 3–6).

It is vital to know your hydrant systems and their capabilities. This should include the grid system and the size of the mains supporting the hydrants. A water main map is a must for this type of operation.

Finally, when using a master stream, the stream itself needs to be moving to cover the entire fire area. Just setting up a master stream and leaving it in one position does nothing more than drill a hole right through the building. This is not an uncommon occurrence on master stream fire operations.

Setting up the operation will require training, but the setup is not overly complicated. When the order is given to employ the super pumper, a predesignated team will go into action. This can be set up in staging with the standard pumpers (fig. 3–7). All units are told to reverse out from the super pumper and go to their designated hydrant. Each unit involved in the relay pumping should take one firefighter to assist in locating the hydrant since you might be looking for a specific one that will provide adequate performance for the operation. It would be a good idea for the pump operator on the super pumper to mark each supply line from the pumpers to indicate which unit the line is connected to for communication purposes.

The pump operator of the super pumper is in charge of the operation. The order should be given for all relays to be charged under idle pressure. Once all

Figure 3–6. Three 6" lines connected from the hydrant to the engine for maximum flow

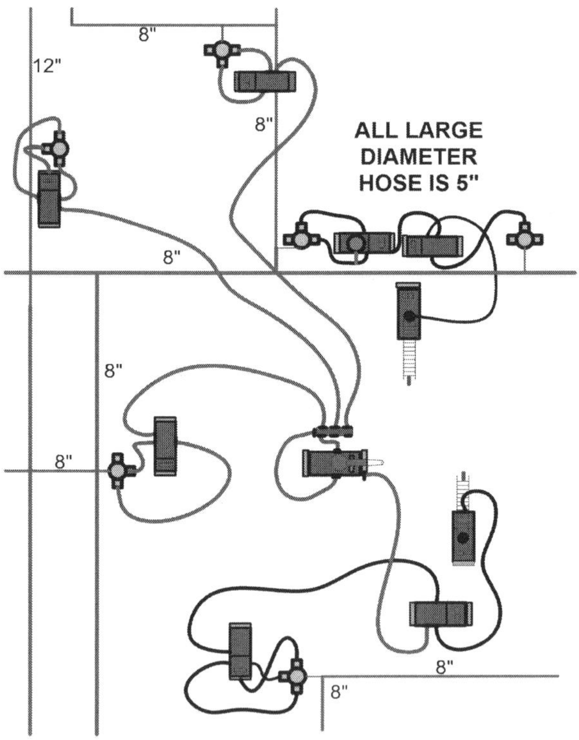

Figure 3–7. This diagram shows an actual fireground water delivery operation, including water mains. The red lines represent super pumper operation. Black lines are normal operations. Blue lines are the water mains.

lines are charged, the order can be given for the pumpers to throttle up (figs. 3–8 and 3–9). Start the operation at about 150 psi, keeping the super pumper at idle and checking to see where the super pumper is at in regard to the flow. If the required flow is close, the super pumper should simply throttle up to make up the difference for reaching the target pump discharge pressure. The goal is to reach the required flow efficiently. In this case, efficient operations will require engine speeds that are not extremely high. Ideally engine speeds will be in the range of 1,700 rpm to 1,800 rpm on all units. The maximum operating pressure of the hose should never be surpassed. Most large diameter hose has an operating pressure of 185 psi, but there is a large diameter attack hose that can go up to 250 psi. Remember, the goal of this operation is twofold. You are trying to reach the desired flow while keeping all pumpers at an efficient level of engine speed vs. pressure, so adjustments may be necessary.

How Many Pumpers Should Be Placed in the Super Pumper Relay?

The main goal to strive for when placing pumpers in a relay is to be able to move the required amount of water in the most efficient manner possible. Efficient operations in this case will mean using as few pumpers as possible.[7]

In large flow relays, the hose becomes a restricting factor in regard to the amount of water that can be delivered. This is primarily due to the maximum allowed operating pressure (mainly with large diameter supply hose at 185 psi) and the length of the hose lay. As a rule of thumb, whenever the flow is doubled, the pressure required to move it is quadrupled. This higher pressure equates to more friction loss in the hose, which in turn has a direct effect on how long the lay can be between pumpers.

A flow chart can be designed to show flow rates and lengths of the line to assist in properly placing the correct number of pumpers in the relay (table 3–2). Our goal is 1,500 gpm per relay.

The bottom line is that the fire service has gotten away from flowing big water. One reason is probably the lack of understanding of the use of large diameter hose. You have to do more than just connect it to the hydrant. The operation in this section is really nothing new. FDNY and others have utilized a similar concept for many years, but some departments have moved away from it. The economy super pumper concept is a way to get back into the big water business with only a small investment, and it has the potential to save millions of dollars in property loss.

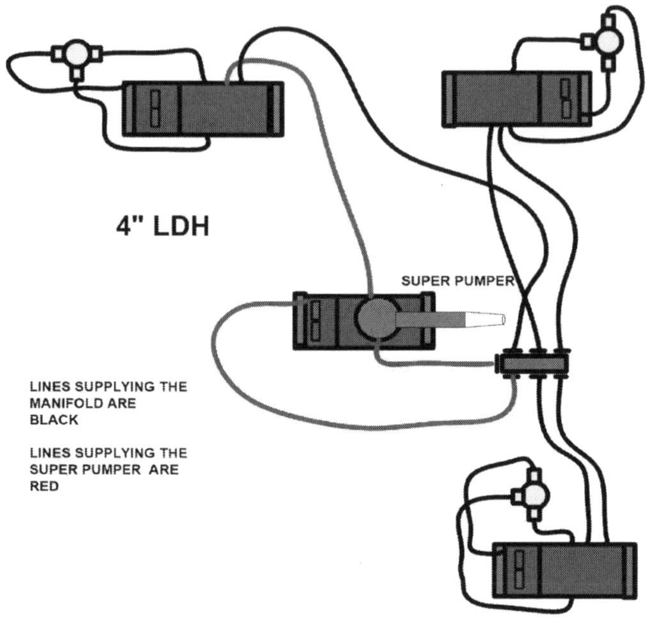

Figure 3–8. A 4" LDH is connected to a super pumper.

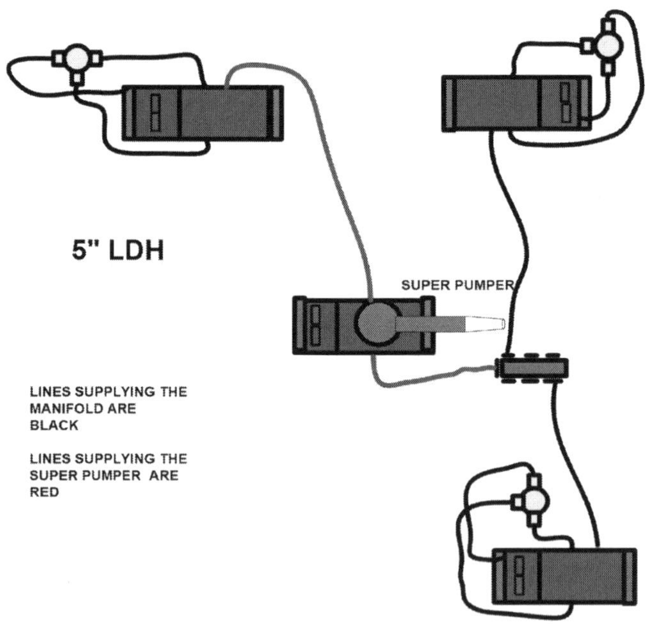

Figure 3–9. A 5" LDH is connected to a super pumper.

Fixed Master Stream Mounted on a Quick Attack Unit

With the quick attack unit, you have to be careful with the water flows because of the lighter weight of the truck compared to a heavier fire truck. In researching pickup truck capabilities, I found it difficult to get anything in writing as to how much nozzle reaction it could handle and still be safe. The truck I used was a 2005 Ford 250 extended cab that weighed 7,800 lbs. I talked with some manufacturers of quick attack units and was informed by one major manufacturer that I could use it to flow 3,000 gpm with the proper installation. They recommended proper connection of the gun to the truck frame and outriggers to support the use of the gun off the side of the truck because of the potential for high nozzle reaction. They also stated that they now use heavier trucks like the Ford 350 and 450 trucks when they want to push the 3,000 gpm mark.[8]

There is only one way to support the quick attack unit flow of 3,000 gpm. It has the same basic principles for the water delivery evolution that the super pumper applies with one exception. The receiving pumper will be supplying the

Table 3–2. Water flows and lengths of line for pumping in relay

4" LDH	
Hose Length	Flow Rate (gpm)
500'	1,500
600'	1,250
1,100'	1,000
5" LDH	
Hose Length	Flow Rate (gpm)
1,500'	1,500
2,200'	1,250
3,300'	1,000

Note: The optimal flow rate is 1,500 gpm. Remember, it takes approximately two 4" lines to equal one 5" line, so if you are using a dual 4" evolution, simply use the data from the 5" chart.

> **Top Tip: Quick Attack Unit**
> Once again, I recommend you allow the pros do the installation work for your Quick Attack Unit project. Many apparatus refurbishing companies offer this service. They employ professional designers and fabricators who know how to build the unit to counteract the high nozzle reaction that will occur from the high flow rates.

quick attack unit. It only has to be a standard pumper, not a super pumper (figs. 3–10 and 3–11).

If you are considering converting one of your engines, you will need the following:

- At least two large diameter intakes with valves
- A minimum of a 3,000 gpm deck gun and tip
- Replacement of the 3" plumbing and valve to the 4" deck gun with 4" plumbing and valve
- Proper bracing for the gun to manage the nozzle reaction
- Large diameter manifold with seven inlets/outlets
- Funding for the conversion (approximately $8,000 based on list pricing from the manufacturers). *Note:* This cost will be less if you are ordering the big gun when you purchase the engine because you will not be buying the small gun. It could even be a wash.

Flow Tests

Bullhead City, AZ, Fire Department big water unit flow tests

With the assistance of the Bullhead, AZ, Fire Department, we conducted a series of flow tests with the goal of achieving a maximum flow of 3,000 gpm from a big water unit using a 4" Stang Master Stream with a 3¼" smoothbore tip. The unit itself was a 2005 Ford 250 XL that weighed 3,800 lbs. Outriggers with a capacity of 5,000 lbs were fabricated and then mounted in the truck trailer hitch receiver.[9]

The tests used three hot hydrants in a commercial area that started with a static pressure of 120 psi. After flowing 3,000 gpm, they had a residual pressure of 30 psi. Each of the three engines used had their own hydrant that they double tapped with 5" and 3" hose. The 3,000 gpm flow was reached with a nozzle pressure of 95 psi (fig. 3–12).

Figure 3–13 shows the engine that pumped the big water unit. Originally it was going to pump dual 5" lines, but the evolution had to be changed because the

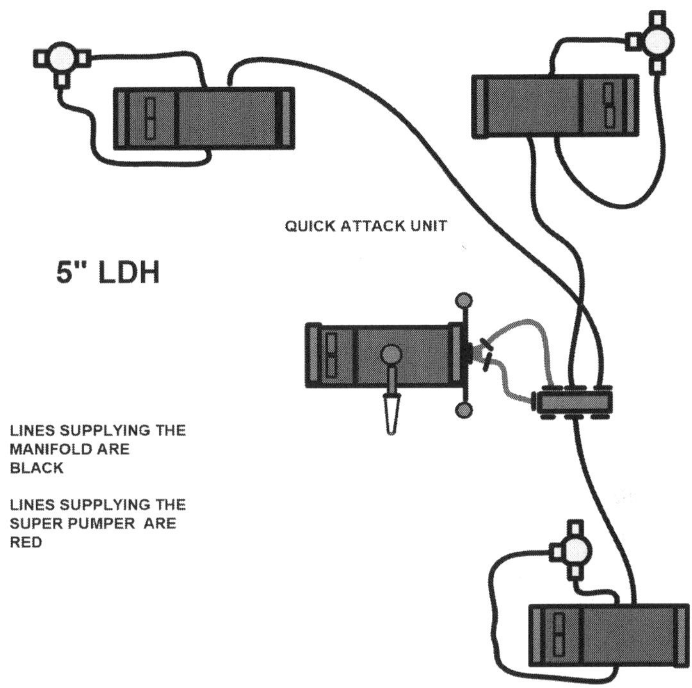

Figure 3–10. This pumper needs to use the high-pressure 5" attack hose. A high-pressure 5" attack hose supports a quick attack unit flow of 3,000 gpm.

Figure 3–11. This master stream is being supplied by two 1,500 gpm pumpers and is flowing 2,451 gpm. It has a reach of 220' using a 3¼" tip at 61 psi nozzle pressure, pushing back a nozzle reaction of 1,012 lbs.

engine did not have enough LDH intakes to receive water from the water supply engines. Thus the plan was altered to allow two engines to pump the big water unit, the original one next to it, and the engine 500' away sitting on another hydrant north of the simulated fireground (figs. 3–14 through 3–17).

Remember to apply the emergency brakes and wheel chocks. I would not recommend flows higher than 2,000 gpm unless otherwise proven to be safe by all manufacturers involved with the project. This type of unit is commonly used with refinery fire departments, so consulting with them could be helpful. In fact, you may want to give Chicago FD a call. Fire departments are there for our protection and should be able to handle to the best of their ability any fire that our community might experience. I know this is a huge responsibility and is

(a)

(b)

Figure 3–12. Bullhead City Fire Department big water flow tests sometimes not totally achievable. Large flow fires, as rare as they are, can and do happen in most communities. For a small investment, the two units covered in this section can provide a big assistance in developing an overwhelming fire stream as compared to the standard master stream flows we are used to. They can provide a big knockdown and still operate on an everyday basis for more common fire responses.

Figure 3–13. Bullhead City FD engine pumping the big water unit

Figure 3–14. Big water unit

Figure 3–15. Notice the volume or thickness of this 3,000 gpm stream.

Figure 3–16. Keep in mind that even though the stream went 200', in reality the target will probably be half of that.

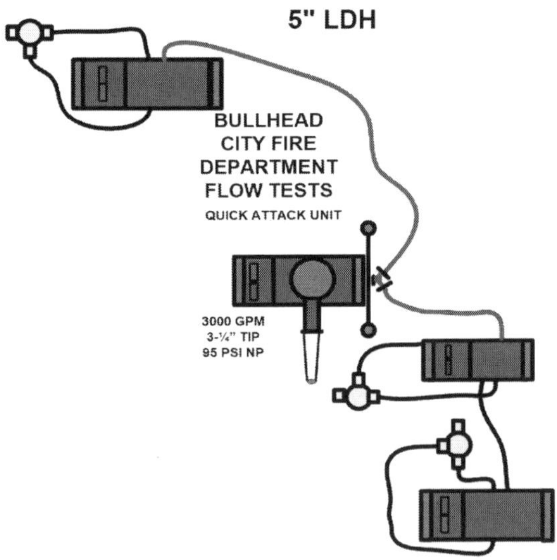

Figure 3–17. Flow test for Bullhead City FD

4

ANCHOR AND HOLD WATER DELIVERY EVOLUTIONS FOR STRUCTURE FIRES IN THE WILDLAND–URBAN INTERFACE

Introduction

This chapter details the water delivery capabilities of anchor and hold structural protection and extinguishment operations in the wildland-urban interface (WUI). In this context, *structure protection* involves actual fire suppression of involved structures, as well as the predeployment of units at structures to defend against an approaching fire (fig. 4–1).

According to FIRESCOPE, a California group that plans for major disasters, including wildland fires,

- "Anchor and Hold" is a tactic utilizing control lines and large water streams from fixed water supplies in an attempt to stop fire spread. The goal is to extinguish structure fires, protect exposures, and reduce ember production.

Figure 4–1. Fire suppression and predeploying units are critical in defending against fire before it reaches structures in the wildland–urban interface.

Photo courtesy of Alan Simmons Productions.

- Anchor and Hold can be referred to as taking a stand to stop the progression of the fire.
- Anchor and Hold tactics are more effective in urban neighborhoods where the fire is spreading from house to house.
- Establishing an anchor and hold line requires considerable planning and effort and utilizes both fixed and mobile resources:
  - Fixed engines should be spotted in safe areas where they can safely withstand any fire situation.
  - Mobile engines or task forces can engage in individual structure defense actions or perimeter control and resupply from fixed water sources.
  - Mobile engines should be prepared to re-deploy to other areas should the fire escape the anchor and hold line.
- Ground resources, such as engine crews and hand crews, should staff hose lines and be prepared to extinguish hot spots, fire perimeters, and structures.
  - Hand crew strike teams should be deployed to construct fire control lines wherever needed and conduct firing operations.

Factors to Consider in Anchor and Hold Strategies

1. Can it be done safely?
2. Is there an adequate water supply?
3. Is there adequate manpower?
4. Is there adequate equipment?
5. Is there a will to get the job done?

The anchor and hold strategies in this chapter are based on five factors: safety, water supply, manpower, equipment, and level of commitment. Two theories or strategies of operation for engine companies protecting structures in the WUI are bump and run (B&R) and anchor and hold (A&H).

Bump and Run

First let's review the *bump and run* tactics. Bump and run tactics are based on the attack unit being mobile and self-sufficient. It relies on its own onboard water (booster tank) for its water supply. This operation is designed to be highly mobile and capable of rapid departure if the situation becomes too dangerous.

With the real prospect of several structures burning in specific areas at the same time, the extreme radiant heat created can be more intense than the heat from burning brush and embers. This factor, along with the possibility of high winds, has the potential for creating firestorm conditions that will require high-flow streams for suppression. When fire enters into the neighborhood, it is no longer a wildland fire, and it can quickly turn into an urban conflagration (fig. 4–2).

When using bump and run tactics, units operate from onboard booster tanks, which limits the water delivery rate capabilities. In order to stretch the mobile water supply, firefighters tend to produce low flow streams, which in turn can be ineffective in knocking down the fire (fig. 4–3). The firefighters go into "preserving water mode" with good intentions but typically wind up wasting water as the inferior streams may lack the muscle to knock the fire down.

The bigger the booster tank is, the longer the water supply will last. In reality, however, many engines tasked with structure protection in the WUI are municipal or city units with 500 to 750 gallon tanks. This water supply can be quickly depleted in a bump and run operation (fig. 4–4).

Anchor and Hold

Tactics

With *anchor and hold* tactics, fire suppression units connect to a hydrant to maintain an uninterrupted water supply, hopefully allowing higher fire stream flow rates.

Figure 4–2. These structure fires were ignited from other structure fires, not burning brush.

Photo courtesy of Alan Simmons Productions.

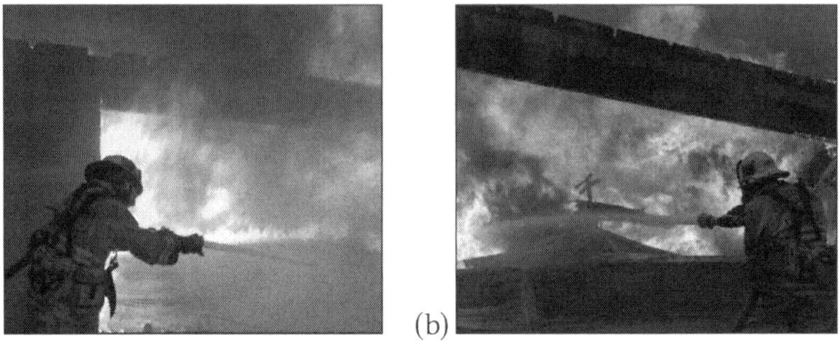

(a) (b)

Figure 4–3. This is what happens in bump and run with booster tanks only! This stream is no match for the flames. You have to overwhelm the heat energy with the necessary water flow rate.

Photos courtesy of Alan Simmons Productions.

Figure 4–4. Small handlines can be used off the booster tank on the engine to hit small spot fires.

Photo courtesy of Alan Simmons Productions.

Retired Battalion Chief Mike Alder from the San Bernardino City Fire Department inspired me to write about the practice. After the Panorama fire, a devastating wildland fire that burned hundreds of homes in 1980, Chief Alder saw the need for a change in tactics to fight this type of fire. He and other department members developed the anchor and hold tactics, which proved to be highly effective in combating the 2003 Old Fire. In their estimation, the anchor and hold tactics saved hundreds of structures. In conducting my research, I met Chief Alder and obtained his permission to seek improved water delivery capabilities for anchor and hold tactics.

The anchor and hold tactics covered in this text were developed using three phases of operations, keeping safety and maximum efficiency as the goals. They are as follows:

1. Maximizing the water supply
2. Stream performance
3. Knockdown capabilities

Fire attack

The basic firefighting strategy for A&H operations is to establish a water supply and initiate an attack on the fire to achieve a quick knockdown on the involved structures to stop or slow down the threat of fire spread to exposures. The safest and most efficient way to accomplish this is with a blitz attack operation. A blitz attack involves hitting a body of fire with an overwhelming amount of water to achieve a knockdown as quickly as possible.

Fire scenarios can vary and thus may require different flows delivered from handlines and master streams. Based on the blitz attack concept as well as the type of tactics involved with fighting these fires, for the most part fire streams will be directed from the outside with the goal of getting quick knockdowns. Simply stated, the higher the flow, the quicker the knockdown.

There are two weapons of choice for this operation when it comes to the attack unit. One is a standard engine company with at least a 3" deck gun, which is the most common size. Keep in mind that this gun can be replaced with a 2,000 gpm gun fairly easily and at a low cost. This configuration will allow for potentially higher flows if available, but more importantly, it allows higher nozzle pressures because of the ability of the higher capacity gun to handle 1,010 lbs of nozzle reaction vs. 631 lbs nozzle reaction with the standard 1,250 gpm. These high-pressure streams (150 psi works best) work well to produce better penetration, as well as being able to perform more effectively in the wind conditions that are almost always present at wildland fires.

The second weapon for this operation is a 4" master stream mounted to a pickup truck with stacked tips ranging from 1⅜" to 2" producing flows from 866 gpm to 1,455 gpm with 150 psi nozzle pressure (fig. 4–5).

The handlines are designed for large flow and are lighter in weight, using 2" hose with 2½" couplings. The flows from the 2" handlines range from 250 gpm to 325 gpm. Again, this will be based on ease of deployment because of the weight of the hose when charged. It may also be based on the handline tactics that will be used. One firefighter is all that will be needed to operate this line. Finally, for

Figure 4–5. A 4" Stang gun mounted to a pickup is fed by a pumper. The flow is 1,455 gpm with a 2" tip at 150 psi nozzle pressure, and knockdown times of 10 seconds are possible. If you look about halfway down the stream, that is where you can expect contact with the fire.

the big hit from a handline the Big Paulie nozzle can be used from a 2½" handline to get a maximum flow of 500 gpm with a 1½" tip at 55 psi nozzle pressure.

Water supply

Fire hydrants represent the most commonly used water source as they are often available on the WUI fireground within the developed structure area. Flow capabilities from the hydrant system can range from extremely poor or out of service to moderate and even high flows.

An engine company will always be placed at the hydrant to set up a relay operation to obtain the most pressure and water from the hydrant. Before deploying the pumping apparatus and the supporting hose evolutions, it is helpful to know how much water is available for the operation. Hydrant use requires a method for evaluating available water. Will the hydrant support a master stream up to 1,000 gpm? Is there more or less than 1,000 gpm? This information could change tactics or even cancel the operation. A quick and realistic flow test can be conducted to help determine available water.

The easiest way to determine an approximate hydrant flow capability is to actually connect the engine to the hydrant based on a high-volume hookup. Then simply fire up the deck gun and take a reading with a handheld *pitot gauge*, using a flow chart to reference the flow (figs. 4–6 and 4–7). A pitot gauge, also called a *pitot tube*, "is a flow measurement device used to measure fluid flow velocity. . . . [It] consists of a tube pointing directly into the fluid flow. As this tube contains no fluid, a pressure can be measured."

Nozzle Pressure in psi*	Nozzle Diameter in Inches																			
	1/4"	3/8"	7/16"	1/2"	5/8"	3/4"	7/8"	15/16"	1.0"	1 1/8"	1 1/4"	1 3/8"	1 1/2"	1 5/8"	1 3/4"	1 7/8"	2.0"	2 1/4"	2 1/2"	3.0"
	Gallons Per Minute																			
20	8	18	25	33	51	75	101	115	132	167	206	250	298	350	407	468	532	674	832	1200
22	8	19	26	34	54	79	105	122	139	175	216	263	313	367	427	490	557	707	872	1255
24	8	20	27	36	56	82	110	127	145	183	226	275	327	384	446	512	582	739	910	1311
26	9	21	29	37	59	85	115	133	151	191	235	286	340	400	464	533	606	769	949	1366
28	9	21	30	39	61	89	119	138	157	198	244	297	353	415	481	554	629	799	985	1418
30	10	22	31	40	63	92	123	142	162	205	253	307	365	429	498	572	651	826	1017	1470
32	10	23	32	41	65	95	127	147	167	212	261	317	377	443	514	591	673	854	1050	1515
34	11	23	33	43	67	98	131	152	172	218	269	327	389	457	530	610	693	880	1082	1561
36	11	24	34	44	69	100	135	156	177	224	277	336	400	470	546	627	713	905	1114	1610
38	11	25	35	45	71	103	138	160	182	231	285	345	411	483	561	645	733	930	1144	1650
40	11	26	35	46	73	106	142	165	187	237	292	354	422	496	575	661	752	954	1174	1692
42	11	26	36	47	74	109	146	169	192	243	299	363	432	508	589	678	770	978	1203	1737
44	12	27	37	49	76	111	149	173	196	248	306	372	442	520	603	694	788	1000	1234	1776
46	12	28	38	50	78	114	152	176	200	254	313	380	452	531	617	710	806	1021	1259	1816
48	12	28	39	51	80	116	156	180	205	259	320	388	462	543	630	725	824	1043	1286	1854
50	13	29	40	52	81	118	159	184	209	265	326	396	472	554	643	740	841	1065	1313	1895
52	13	29	40	53	82	121	162	188	213	270	333	404	481	565	656	754	857	1087	1339	1930
54	13	30	41	54	84	123	165	191	217	275	339	412	490	576	668	769	873	1108	1364	1968
56	13	30	42	55	86	125	168	195	221	280	345	419	499	586	680	782	889	1129	1389	2000
58	13	31	43	56	87	128	171	198	225	285	351	426	508	596	692	796	905	1149	1417	2020
60	14	31	43	57	89	130	174	201	229	290	357	434	517	607	704	810	920	1167	1438	2075
62	14	32	44	58	90	132	177	205	233	295	363	441	525	617	716	823	936	1187	1463	2108
64	14	32	45	59	92	134	180	208	237	299	369	448	533	627	727	836	951	1206	1490	2140
66	14	33	46	60	93	136	182	211	240	304	375	455	542	636	738	850	965	1224	1510	2175
68	14	33	46	60	95	138	185	214	244	308	381	462	550	646	750	862	980	1242	1532	2205
70	15	34	47	61	96	140	188	218	247	313	386	469	558	655	761	875	994	1260	1558	2240
72	15	34	48	62	97	142	191	221	251	318	391	475	566	665	771	887	1008	1278	1579	2275
74	15	35	48	63	99	144	193	224	254	322	397	482	574	674	782	900	1023	1296	1600	2300
76	15	35	49	64	100	146	196	227	258	326	402	488	582	683	792	911	1036	1313	1620	2340
78	15	36	50	65	101	148	198	230	261	330	407	494	589	692	803	924	1050	1330	1640	2370
80	16	36	50	66	103	150	201	233	265	335	413	500	596	700	813	935	1063	1347	1665	2400
82	16	37	51	66	104	152	204	235	268	339	418	507	604	709	823	946	1076	1364	1685	2425
84	16	37	51	67	105	154	206	238	271	343	423	513	611	718	833	959	1089	1380	1707	2460
86	16	37	52	68	107	155	208	241	274	347	428	519	618	726	843	970	1102	1396	1723	2485
88	16	38	53	69	108	157	211	244	277	351	433	525	626	735	853	981	1115	1412	1745	2515
90	17	39	53	70	109	159	213	248	280	355	438	531	633	743	862	992	1128	1429	1762	2540
92	17	39	54	70	110	161	215	249	283	359	443	537	640	751	872	1002	1140	1445	1785	2565
94	17	39	54	71	111	162	218	252	286	363	447	543	647	759	881	1012	1152	1460	1802	2600
96	17	40	55	72	113	164	220	255	289	367	452	549	654	767	890	1022	1164	1476	1822	2625
98	17	40	55	73	114	166	223	257	292	370	456	554	660	775	900	1032	1176	1491	1840	2650
100	18	41	56	73	115	168	225	260	295	374	461	560	667	783	909	1043	1189	1506	1860	2680
105	18	42	57	75	118	172	230	266	303	383	473	574	683	803	932	1070	1218	1542	1908	2745
110	19	43	59	77	121	176	236	273	310	392	484	588	699	822	954	1095	1247	1579	1950	2810
115	19	43	60	79	123	180	241	279	317	401	495	600	715	840	975	1120	1275	1615	1995	2875
120	19	44	61	80	126	183	246	285	324	410	505	613	730	858	996	1144	1303	1649	2020	2940
125	20	45	63	82	129	187	251	291	331	418	516	626	745	876	1016	1168	1329	1683	2080	3000
130	20	46	64	84	131	191	256	295	337	427	526	638	760	893	1036	1191	1356	1717	2120	3055
135	21	47	65	85	134	195	262	302	343	435	536	650	775	910	1056	1213	1382	1750	2160	3110
140	21	48	66	87	136	198	266	308	350	443	546	662	789	927	1076	1235	1407	1780	2200	3170
145	21	49	68	88	139	202	271	313	356	450	556	674	803	944	1095	1257	1432	1812	2240	3225
150	22	50	69	90	141	205	275	318	362	458	565	686	817	960	1114	1279	1456	1843	2260	3280
175	24	55	74	97	152	218	298	344	390	494	610	740	883	1030	1200	1380	1572	1985	2460	3540
200	26	59	79	104	162	234	318	368	418	527	653	790	944	1103	1283	1472	1680	2123	2635	3785

* Nozzle pressure measured by pitot tube and gauge

Figure 4–6. A flow rate reference chart

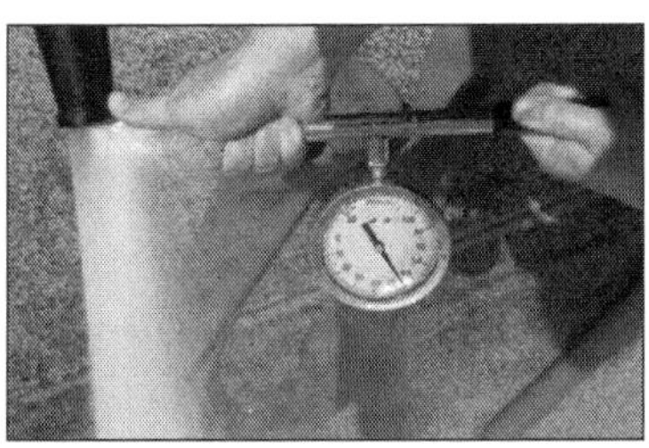

Figure 4–7. Handheld pitot gauge

A hydrant worth using should hopefully have a flow rate of at least 1,000 gpm. If you are only able to flow 500 gpm or less, it may be questionable whether or not the operation can be supported from this hydrant. The choice would be yours.

Water Supply Hose Evolution with a Pumper or Quick Attack Unit

In the early days of anchor and hold tactics, a common hose evolution involved a small diameter hose supply line connected directly from the hydrant to support pumpers attacking the fire. This concept dramatically decreases the flow from the hydrant because of the higher friction loss in the 2½" or 3" supply lines, as well as the low fixed system pressure available from the hydrant (figs. 4–8 and 4–9).

There are several water delivery hose evolutions that can be used to move water in an interface operation more efficiently than the above-mentioned operation. In this section we will focus on the use of 4" and 5" large diameter hose for the operations. Often equipment such as supply hose is unofficially standardized for certain suburbs for multiple reasons. One reason is that it allows mutual aid companies to easily work together. A good example is the Southern California fire service, which is known to have major interface fires and uses predominantly 4" hose for supply lines. One reason is that 4" is smaller in size and is theoretically easier to run over in regard to clearance. While this is true for automobiles, in my experience I have never seen a fire apparatus that could not clear 5" large diameter hose. Do not run over the couplings of any size hose, however, since it would destroy the couplings.

The new evolutions described in this section are based on using only two units per evolution. This is done for two reasons. The first is that because we are trying to achieve a big knockdown on a structure fire, we are looking to achieve at least a 1,000 gpm stream, and hopefully higher. A single hydrant in a residential area will most likely not have much more than that, so using multiple pumpers at the same time will interfere with the goal of a 1,000 gpm stream. The other reason is that traditionally on wildland fires, units are scarce. Multiple units on a single evolution may not be possible.

I recommend using 5" large diameter attack hose for the operation simply because it has a higher operating pressure (275 psi) than the large diameter supply hose, which can only be used up to 185 psi. The higher pressure hose will allow the following operations to be conducted with better results:

- *High-pressure smoothbore tip operations at a maximum of 150 psi nozzle pressure.* This increases penetration and helps battle the stream-destroying winds.

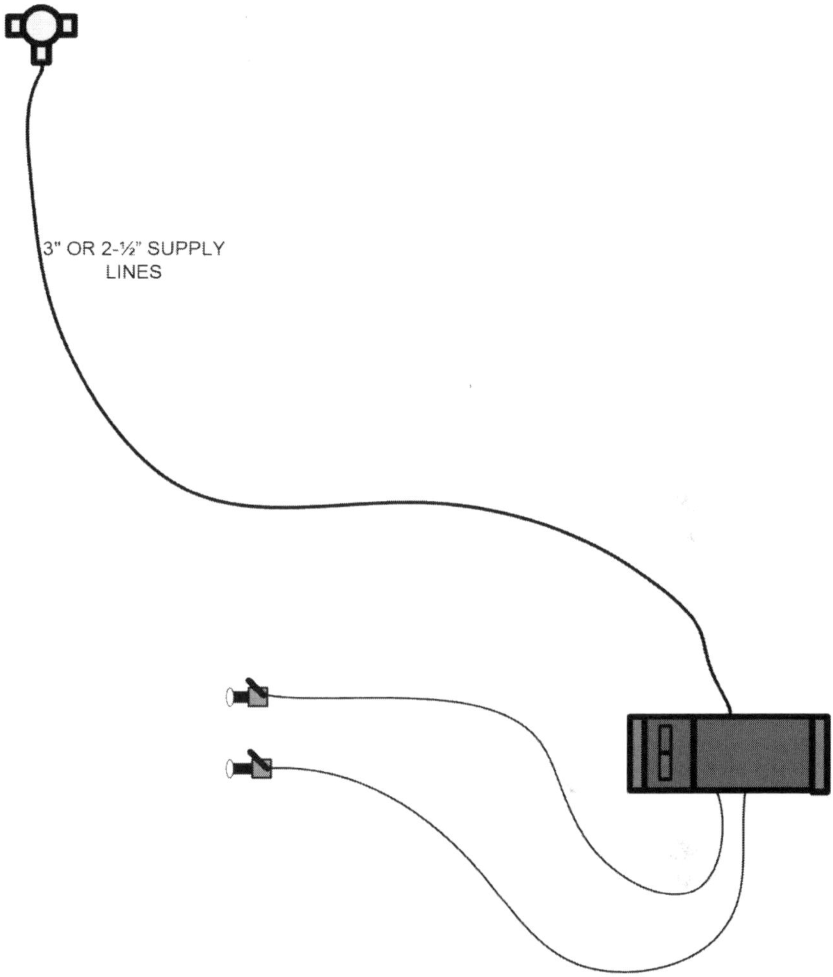

Figure 4–8. A single 2½" or 3" supply line connected directly to the hydrant

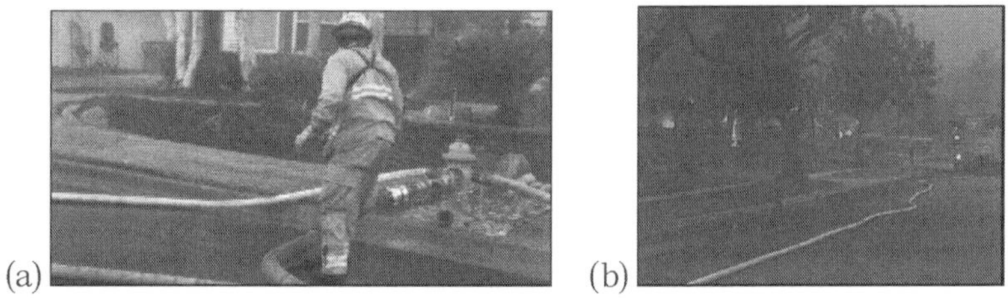

Figure 4–9. This hydrant probably had a capability of about 1,200 gpm, but the flow was stifled because of the size of the supply hose and the lack of a pump on the hydrant to boost the pressure.

Photos courtesy of Alan Simmons Productions.

- *Longer supply lines because of less friction loss.* Basically, a 5" line can move twice as much water as a 4 line, as long as the water volume is available from the hydrant.
- *Greater output from the 5" line.* Based on the water supply, the 5" line has a better chance of supplying more than one unit than the 4" line.

Even though the 4" and 5" lines have much lower friction loss than the 2½" and 3" lines, connecting them directly to a hydrant using only hydrant pressure is not optimal. It will not allow the system pressure to move the maximum available water through hose evolutions that stretch any distance at all from the hydrant. As mentioned previously, the most efficient method for moving the required flow through the hose evolution is to place a pumper at the hydrant in a relay pump operation to discharge the water under higher pressure to the receiving unit (fig. 4–10). The unit at the hydrant should make its hookup with at least two LDH soft suctions. If the hydrant seems to be hot enough, add a third soft suction if there is a means to do so. The attack unit, whether it is another engine or an attack unit such as the one featured in this section, will then lay a single line down the street that the fires are on. If multiple targets will be covered by the evolutions, the operations may employ jumbo gated wyes, or large hose appliances with gates that can be opened or shut to allow or prevent water flow from either of the two discharges. The jumbo gated wyes should be placed in the

Figure 4–10. An engine should hook up big when pumping from the hydrant in an anchor and hold operation.

Photo courtesy of Alan Simmons Productions.

evolution at specified locations to allow the single pumper to connect where needed without shutting the operation down.

If there are more pumpers available, another option is to allow these units to make the hookup at one of the large wyes as opposed to moving a single pumper from one wye to another. This will allow for a quicker knockdown, but it is important that none of the operators involved with the evolution steal water from the unit flowing the big water from the master stream. Another task that an extra engine can do is pump from another hydrant into one of the jumbo manifolds and boost the available water. There is a good chance of nearly doubling the flow if the hydrants are working.

Any water in excess of the flow being produced by the main master stream attack will still be transferred to other pumpers if available in the evolution. This could allow for small handline operations until it is their turn to flow the bigger stream.

The quick attack unit mentioned can be used in place of the attack pumper using two similar evolutions with the supply pumper at the hydrant. Because there are no pumps on the quick attack units, the water will have to be used by a single quick attack unit at a time for the big flows. Even though the engine at the hydrant is pumping at maximum pressure, it not only has to overcome the friction loss in the hose, it must also develop the nozzle pressure as well.

The method for employing the quick attack unit involves only one quick attack unit and the engine at the hydrant. When the relay line is laid, a jumbo gated wye can be connected in-line where a fire attack is needed and left open on the relay discharge port to provide the water to the entire relay evolution. Several additional jumbo gated wyes can be connected in-line where needed and will also be left open for the relay. The quick attack unit can start to work at any of the wyes based on the situational needs. When a knockdown is achieved, the crew simply closes the gate on the wye supplying the quick attack unit and goes to the next one and makes the hookup again. If needed, a smaller gated wye can be connected to any of the jumbo wyes that are not in use by the quick attack unit to supply small handlines.

It is a known fact that the fireground area in the wildland interface and conflagration can be extremely dangerous, thus requiring the rapid exit of the area to maintain safety. Because supply lines are being used in anchor and hold operations, a quick disconnect is crucial (fig. 4–11). The quickest way to do this is to have the jumbo gated valve within the supply line evolution close to the pumper. Connect a short section of large diameter hose about 15" between the gate valve and the pumper intake (fig. 4–12). When the valve is shut down on the wye, excess pressure from the intake valve can be drained in seconds.

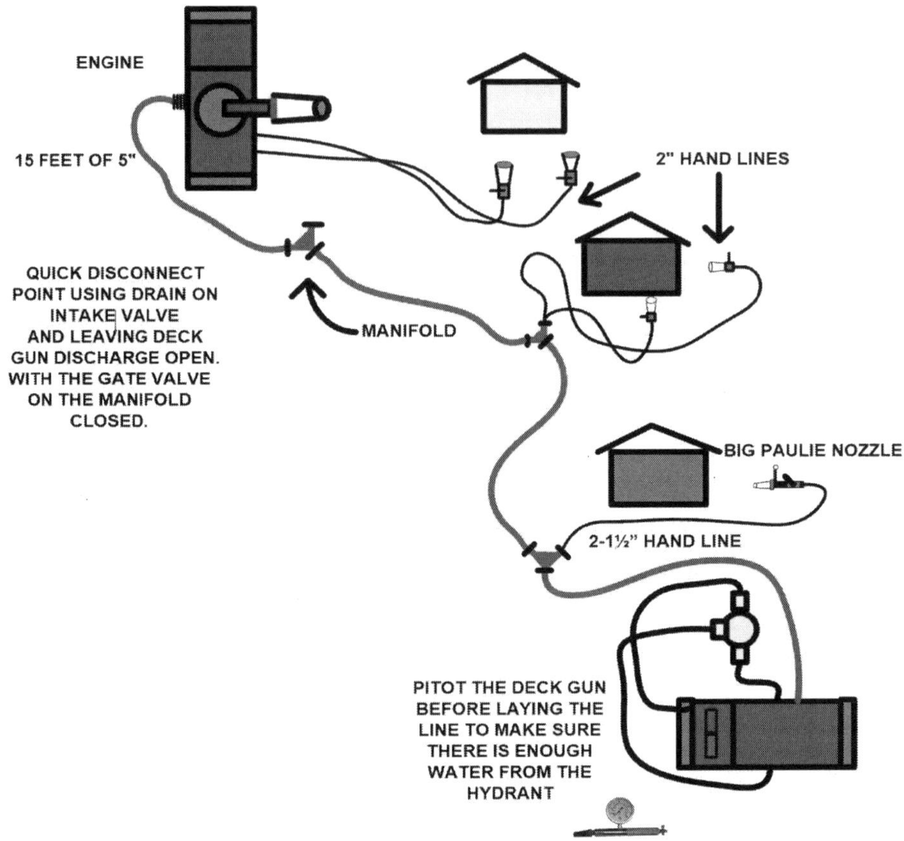

Figure 4–11. A quick disconnect is important because supply lines are being used in anchor and hold operations.

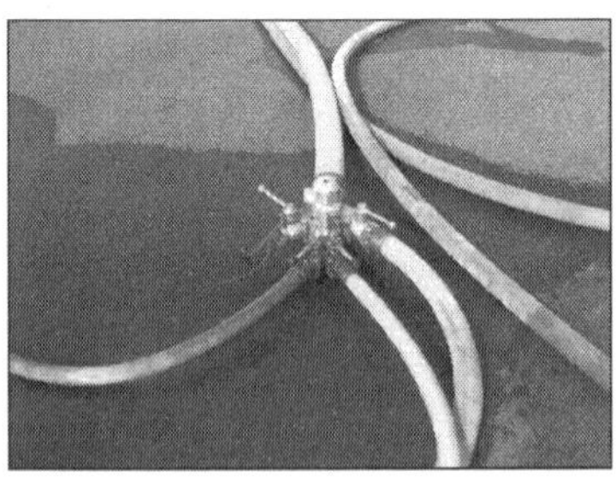

Figure 4–12. High-flow handlines can be connected to the manifold to add more streams or hit targets that the master stream cannot reach.

Handlines

The handlines in an anchor and hold operation are designed for three different target flows. I say "target" because it could be difficult to dial in the exact flow rate. You want to be able to flow a big waterline for the areas that the deck gun cannot reach (fig. 4–13). A flow of 265 gpm is suggested and possible. You want to be able to flow an interior attack flow (not that you will go interior), and 150 gpm works well. Finally you want a mop-up line flow, usually about 30 gpm (the flow is not critical), which can be achieved by either adjusting the pressure or simply gating the nozzle down.

To make this operation as simple as possible, one size of handline is used to obtain all flows. It is 2" hose with 2½" couplings. This hose is just a couple of pounds heavier than 1¾" line, and its maneuverability is really not an issue because the lines are not likely to be used for the interior attack. Of course, a 2½" handline can be used with the Big Paulie nozzle to flow 500 gpm.

These lines are connected to any of the manifolds in the evolution or to the attack pumper if used. Dialing in the correct discharge pressure is done at the manifold or at the pumper if used. A simple reference chart should be designed to show the friction loss for each flow based on a 50" section of hose and the type of nozzle being used. The nozzle pressure should also be added.

It can be frustrating to choose the correct nozzle pressure. You want a fairly high nozzle pressure when the wind is kicking up to ensure you can hit the target. On the other hand, you may need low nozzle pressures when the evolution lacks sufficient pressure to flow the lines at the proper flow rates. Just remember that

Figure 4–13. "Big Paulie" nozzle, City of Los Angeles–style, for the areas the deck gun cannot reach

Photo courtesy of Alan Simmons Productions.

most nozzles can be gated down if the pressure is too high and still produce an effective stream. The most efficient technique in this operation will likely be to sit on the line, which will also help absorb higher nozzle reactions. Table 4–1 offers examples of nozzle pressures and flow rates you might expect from an anchor and hold evolution. These flows are based on friction losses for 50' sections of hose.

Since it will not be possible to set the discharge pressure for the handline from the manifold by the pumper, it has to be done at the manifold itself. An inline gauge should be placed between the discharge port of the manifold and the handle, and the pressure should be set while the line is flowing (fig. 4–14).

Table 4–1. Nozzle sizes and pressures and accompanying flow rates and friction losses for a typical anchor and hold evolution

Hose Size	Nozzle Size/Tip	Nozzle Pressure (psi)	Flow Rate (gpm)	Friction Loss (psi) per 50'
2"	1⅛" Smooth bore	50	265	10
2"	Fized-Gallonage	50	265	10
2½"	1¼" Smooth bore	50	325	11
2½"	1½" Smooth bore	55	500	25

Figure 4–14. A 2½" × 2½" gated wye with gauge mounted on the jumbo wye itself, as well as connected to the outlet before the hose

Fixed Master Stream

The main purpose of the master stream is to quickly knock down the fire at a burning structure (fig. 4–15). Water priority should be given to the master stream. When used effectively, the master stream can achieve a knockdown in a minute or less. High-pressure smoothbore tip operations are encouraged for deep penetration and to combat wind conditions. The effective flow range should be between 500 and 1,500 gpm. If the master stream appliance in use is capable of 2,000 gpm or higher, and sufficient water is available, a flow rate of 1,455 gpm at 150 psi nozzle pressure can be produced safely (table 4–2).

High pressure streams are based on two rules set by the manufacturer. The first is the maximum inlet pressure for the gun, which is 150 psi to 200 psi. The second is the maximum allowed nozzle reaction for the gun, which is 1,010 lbs for a 2,000 gpm gun.

I recommend installing a gauge on the appliance that can be used as a reference for dialing in the correct nozzle pressure and pump pressure from the water source pumper. Understand that the gauge will probably not be accurate for finding the nozzle pressure because of its location. It is only a reference point that the firefighter can check to know when the desired pressure is reached.

Example procedure

The first step is to conduct a flow test with a handheld pitot gauge to get 150 psi nozzle pressure on whatever size tip you are using. Once you reach the desired number, look at the gauge on the appliance and note the pressure reading. This

Figure 4–15. Master streams are designed to quickly knock down fire at a burning structure.

Photo courtesy of Alan Simmons Productions.

Table 4–2. Pump discharge pressure chart for master stream and handlines

Maximum Inlet Pressures to Gun and Nozzle Reactions (NR)			
AKRON 200 psi	TASK FORCE TIPS 200 psi	ELKHART 200 psi	STANG 150 psi
1,250 gpm Gun 631 lbs NR 2,000 gpm Gun 1,010 lbs NR			
Deck Gun Chart			
Nozzle Size	Nozzle Pressure (psi)	Flow Rate (gpm)	Deck Gun Gauge Pressure (psi)
2"	150	1,455 gpm	160 (target pressure)
1¾"	150	1,114 gpm	
1½"	150	819 gpm	
1⅜"	150	688 gpm	
2" Handline Chart			
Nozzle Size	Nozzle Pressure (psi)	Friction Loss (FL) (psi) for 50'	Flow Rate (gpm)
Fized-Gallonage Combo	50	10	250
1⅛"	50	10	250
1¼"	50	20	325
2½" Handline Chart			
Nozzle Size	Nozzle Pressure (psi)	FL (psi) for 50'	Flow Rate (gpm)
1½"	55	25	500

can now be placed on a reference chart that will tell the firefighter when 150 psi tip pressure is reached. Remember that the gauge will not match the nozzle pressure because of its location.

Flow tests must also be done the same way to figure maximum flows when using a 1,250 gpm gun to ensure that the flows do not break the rules set by the manufacturers.

Firefighting 101 has a simple rule. You have to "match the GPMs with the BTUs." If the goal is to put the fires out, this rule must be followed. The operations discussed can provide a safe and efficient means to get this done as long as the five requirements mentioned at the beginning of this chapter can be met.

It is important to make sure all units are on the same page as far as what evolution will be used. A water management officer should be designated to oversee operations. The person filling this position should be familiar with the evolutions as discussed, available water supply, units involved, and available manpower. Remember that wildland–urban interface fires may involve new or fairly recent developments that could have a designated water management officer.

The anchor and hold operations mentioned in this section require everyone involved to be on the same page. Often strike teams are made up of units from various agencies. These units may have different types of equipment as well as varying operating procedures. For this reason, a preincident briefing should be conducted before implementing any hose evolution, especially the anchor and hold operations discussed in this section.

5

DELIVERING 2,000 GPM FROM AN ELEVATED STREAM

This chapter will examine how to use 2,000 gpm *quints* and other apparatus with 2,000 gpm elevated water delivery capabilities to their fullest potential. A quint is a combination fire service apparatus that functions as "an engine and a ladder truck. The name 'quint'. . . refers to the five functions that a quint provides—pump, water tank, fire hose, aerial device, and ground ladders."[1] The goal is to reliably deliver 2,000 gpm water flow at the fire scene (fig. 5–1).

Figure 5–1. Most elevated platforms are designed to flow 2,000 GPM whether they are Quints or non-Quints.

Introduction

Consider a scenario involving a three-story apartment complex in its last phase of construction. It has a wood frame structure, and although construction is not yet complete, the roof is in place. This structure fire is well involved, with immediate exposure problems on three of the four sides of the complex. The first alarm companies lay in from the two closest hydrants and attempt a quick interior attack to no avail. The chief then orders everybody out, with the goal of fighting the fire with exterior operations deploying heavy streams. Because the closest hydrants are on the same loop and the engines are already using those hydrants, every drop of water from this hydrant system is already in use. As a result, companies are having a hard time deploying heavy streams. The chief then orders a quint with a 2,000 gpm water delivery capability from the platform into service, with the intention of deploying an elevated stream to attempt a knockdown as quickly as possible. Since the chief has been in similar situations, he is not surprised that not only is the platform not capable of producing the desired 2,000 gpm water flow, it has difficulty even producing 1,000 gpm flow. There was not enough water from the hydrant system to support the necessary operations.[2]

In order to fight fire, it is important to have the appropriate tools. So what tools are needed to flow 2,000 gpm through a 2,000 gpm elevated stream? They include the following:

- For apparatus other than quints, attack-hose rated hoseline (due to the high required inlet pressure)
- Hydrant or hydrants capable of 2,000 gpm
- The correct supply line evolution to deliver 2,000 gpm
- A pump capable of delivering 2,000 gpm
- The proper discharge pressure to deliver 2,000 gpm
- The appropriate appliance and nozzle to deliver 2,000 gpm

Hydrants

Does a single fire hydrant have the capability of flowing 2,000 gpm? The answer is, "Maybe." Obviously, a hydrant system that is designed by code in a potential high-flow area would have the best chance of achieving this goal. That does not necessarily mean it will, however.[3]

An important consideration to remember is that the potential of the fire hydrant is not totally based on the static pressure. I am sure you have heard someone say, "Oh yeah, this hydrant is great. It's got a pressure of 80 psi." As explained in chapter 1, the true test of a hydrant is what it is capable of flowing under whatever system pressure it has. The system pressure when flowing is actually the residual pressure. If a hydrant rated at 80 psi is opened up, and after flowing 1,000 gpm the residual pressure comes down to 10 psi, it is not a "hot" hydrant. It is true that the initial pressure was high, but after flowing 1,000 gpm, which is below most code requirements, the hydrant was out of water.

Flow Tests

Returning to our initial goal of 2,000 gpm, should we bypass the hydrant just described, or utilize it and supplement it with more water? Let's say that we choose to utilize it. The next step would be to supplement this hydrant from another hydrant, whether the next one up or possibly another one a little further away that has more potential because it is supplied by a different main system. The following flow tests were conducted on an 8" looped hydrant system utilizing two hydrants similar to the scenario previously described. The second part of the flow test was to set up supply line evolutions and repeat the flow tests.[4]

Fire Hydrant Statistics

- Static pressure—70 psi
- Total capacity of main supplying hydrants in excess of 3,500 gpm
- Flow out of single 2½" hydrant outlet—1,188 gpm
- Flow out of single 4" hydrant outlet—2,016 gpm
- Flow from 2½" and 4" hydrant outlets from single hydrant at same time—2,348 gpm
- Flow from 4" hydrant outlets from two hydrants at same time—3,216 gpm

These four supply-line evolution flow tests showed different scenarios for getting water from a hydrant system (fig. 5–2). The results were interesting, and several results were notable. First notice the difference in flows between the single lay using the 2½" port and the 4" port. The 4" port was only able to flow 250 gpm more than the 2½" port. You may wonder how that could be the case when the hydrant flow tests resulted in 2,016 gpm and 1,188 gpm, respectively. The answer is simple. The 5" hose, large as it is, and the intake valve were the restricting

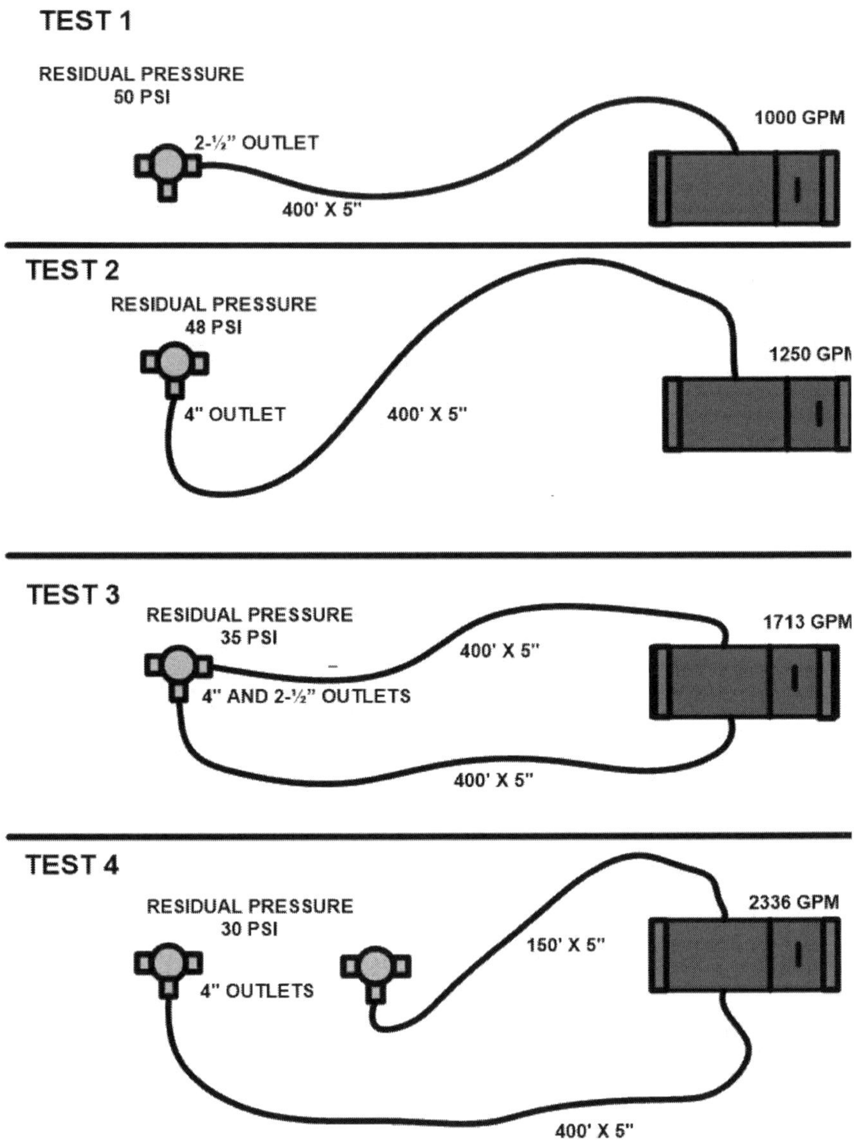

Figure 5–2. Four different ways to get water from a hydrant system

factors. Even though the system started off with a 70 psi pressure before the water was flowing (static pressure), the actual pressure that was available to move the water was lowered to 48 psi and 50 psi, respectively. These flows were good, but they would not allow a pumper to be used to its fullest capacity if necessary.

As you can see, test number four was the only test that was capable of delivering 2,000 gpm. Test number three could have potential on an exceptional hydrant, but I have found in my experience that as a general rule, most hydrants cannot flow 2,000 gpm, even with two supply lines connected to them.

Supply Line Evolutions

There are several supply line evolutions that can be used to support a 2,000 gpm stream. It is important to remember that the evolution design is based on multiple factors, including the potential of the hydrant system, the diameter of the hose, the length of the supply line, and the setup of a *proactive* evolution. By proactive I mean setting up for a worst-case scenario in regard to the hydrant potential. This means placing a pumper on the second hydrant in relay every time to allow for the greatest potential of water delivery. If you roll the dice and set up the evolution without a pump on the second hydrant, and the water delivery is insufficient, you will lose additional time before the evolution can be augmented with a second pumper.[5]

Utilizing the Full Potential from the Hydrant

Quints

Most municipal departments have moved away from using 2½" supply lines. Even so, let's consider what it would take to move 2,000 gpm with 2½" hose and two hydrants with statistics similar to the hydrant flow tests previously discussed (fig. 5–3). Because 2,000 gpm pumpers are not common, while 1,500 gpm pumpers are, the supply line illustrations we will discuss will have 1,500 gpm pumps, and they will be supplying a quint.[6]

As you can see, using 2½" hose is workable. In fact, besides having multiple lines laid to accomplish this feat, the engine pressures are very respectable. Note that the multiple lines coming into the intake seemed to center at one point. This illustration uses the 6" intake on the pump with a trimese 2½" appliance connected to it. A *trimese valve* is a water delivery appliance that combines multiple hoselines into one line, allowing up to three lines to be used. Note that the lines

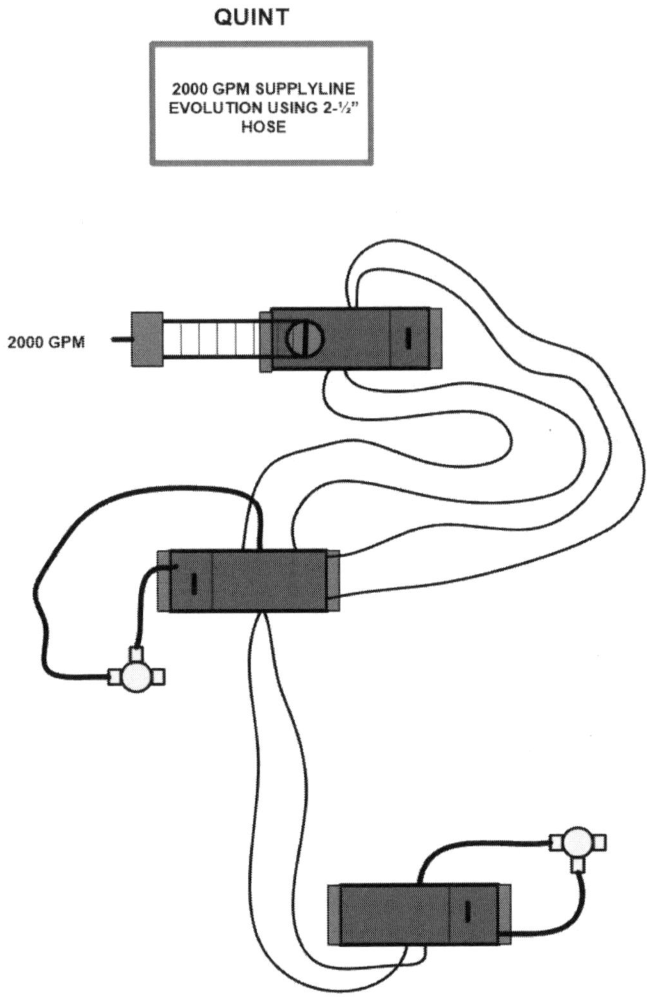

Figure 5–3. This is a diagram of a Quint utilizing 2½" discharge lines.

that connect the engines to the hydrant are thicker, indicating that they are large diameter hose soft suctions.

Next, we will take a look at the same operation with the same goal of moving 2,000 gpm of water, but this time we will use 4" *and* 5" supply hose, with a maximum operating pressure of 185 psi for the supply lines (fig. 5–4). Note that the evolution for each size hose is the same.

The differences are notable. First, we have eliminated the relay operation between the truck company pump and the first hydrant. Second, we have eliminated a total of five lines because of the lower friction loss in the 4" and 5" lines.

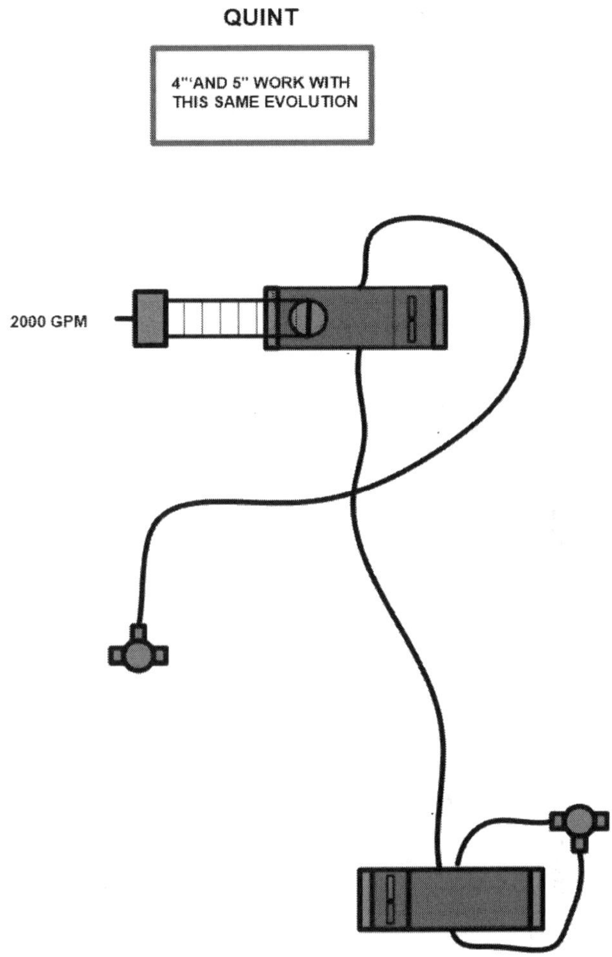

Figure 5–4. The 4" and 5" lines both work with the same evolution.

Notice that the use of two hydrants has not changed. As previously discussed, one hydrant will not usually be able to produce 2,000 gpm, even with multiple lines connected to it.

The 5" operation was able to mirror the 4" operation. It is true that there is a possibility of two hydrants delivering 2,000 gpm total with the 5" line without a relay. Nevertheless, we are being proactive and placing a pumper on the second hydrant to get maximum flow (fig. 5–5). If the relay is able to produce more than 2,000 gpm, there is no harm. On the other hand, if the relay has not been established and our 2,000 gpm goal cannot be met, it will significantly more time and effort to play catch up.

Figure 5–5. This 2,000 gpm quint is being supplemented by a 300' tandem pump operation. The first hydrant was good for about 1,600 gpm, so the tandem operation gave it the extra water plus assisted in the overall pressure and engine speed requirements for the operation.

Other apparatus

There are also elevated platforms that are capable of 2,000 gpm flows that do not have pumps. Let's consider what it would take to supply these trucks using the same hydrant system used for the quints. The discharge line from the pumper supplying the truck company is large diameter attack hose, with a maximum operating pressure of 275 psi (figs. 5–6 and 5–7).

Pump capacity

As mentioned before, all pumps used in the supply line evolution illustrations were rated at 1,500 gpm. You might be wondering how that is possible when the flow was 2,000 gpm for the pumper supplying the quint, as well as the quint itself. The rated capacity of a pump is based on the net engine pressure. *Net engine pressure* is the pressure created entirely from the power plant of the fire apparatus operating the pump. Because of the way pumps are designed, as pressure rises above 150 psi net engine pressure, the capacity starts to drop. So how is it possible to achieve output that is greater than the rated capacity of a pump? When a pressurized water source is introduced into a pump, such as from either a hydrant

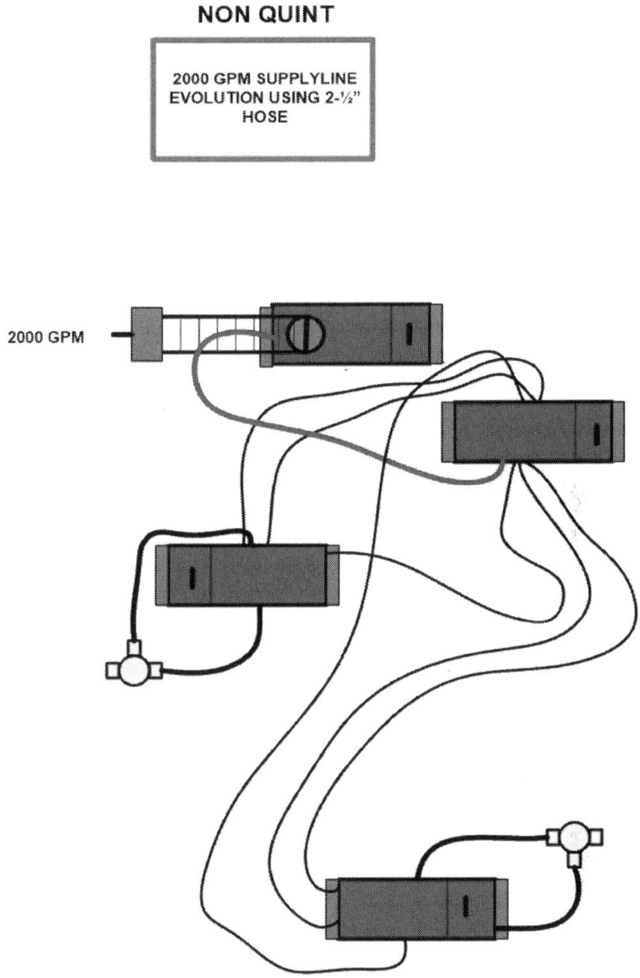

Figure 5–6. A 2,000 gpm supply line evolution using 2½" hose

system or another pump in relay, the intake pressure adds to the net engine pressure. As a result, the incoming pressurized water source reduces the amount of overall work required by the power plant of the apparatus to create the net pressure.

The proper discharge

The most appropriate discharge to deliver 2,000 gpm is the large diameter discharge. Bigger is better in this case, and a 4" discharge with either a 3½" or 4" valve would be the ultimate setup. A 3" discharge could work, depending on its design, but there would be more friction loss. If all you have on your pumper is 2½"

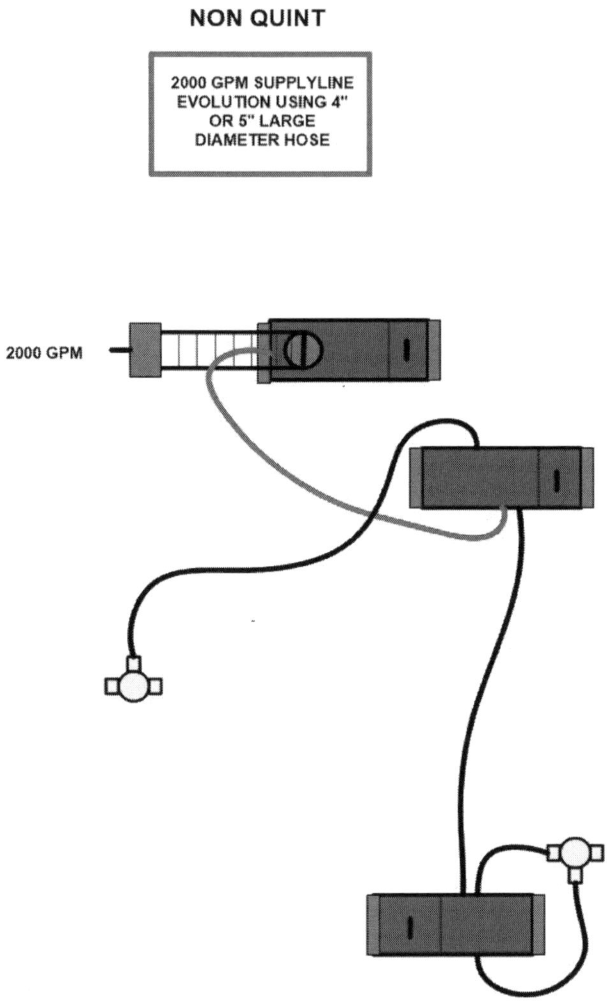

Figure 5–7. A 2,000 gpm supply line evolution using 4" or 5" LDH

discharges, I would recommend using two large diameter hoselines to the quint, simply because a single 2½" discharge can only efficiently deliver around 1,000 gpm. If a 2½" discharge is to be used, make sure it is the one coming off the side of the unit. Rear 2½" discharges have far too much plumbing and therefore cannot deliver even 1,000 gpm.

The required master stream appliance and nozzle

In order to flow 2,000 gpm from a master stream appliance, it has to be rated to do so. The standard master stream appliance, whether it is on an engine company or a truck company, is usually rated at 1,250 gpm. Due to the manufacturer's recommendations, it cannot be used to deliver more than that. Thc good news is that if you have a 2,000 gpm waterway on a quint, it will more than likely have a 2,000 gpm appliance. The most common 2,000 gpm master stream nozzles are the automatic and the smoothbore nozzles. The automatic nozzle typically is designed to flow from around 150 gpm to 2,000 gpm. Nothing changes on the nozzle; it operates automatically. There are two different sizes of smoothbore nozzles that can be used to flow 2,000 gpm. The first one is the 2½" diameter tip. The nozzle pressure needed for the 2,000 gpm flow is approximately 118 psi. The other smoothbore nozzle is a 2¾" tip, with a required nozzle pressure of 80 psi.

It is important to follow all the ladder operation guidelines for the specific unit you are using. Most platforms have a wide range of use in regard to the elevation and angle at 1,250 gpm. The 2,000 gpm flow ranges, however, will be more restrictive because of the higher nozzle reaction that is produced (fig. 5–8). Every ladder truck will have a range capability chart at the pedestal, so make sure to follow what it says.

As you can see, having the proper equipment to deliver the required stream is only half the battle. A complete understanding of everything involved with a pump operation is a must to utilize the equipment to its fullest capacity.

Figure 5–8. Note the difference between the elevated 2,000 gpm stream and the 1,000 gpm deck guns.

6

WATER DELIVERY OPERATIONS FOR FIRES IN HIGH-RISE STRUCTURES

Introduction

Fighting a high-rise fire is challenging because it requires firefighters to develop and deploy unique strategies and tactics. Obtaining the required flow may be the most important issue for high-rise firefighting. The pressure restriction problems common in fire protection systems have challenged firefighters to develop new tactics to achieve the required streams. For the most part this has been accomplished by using high-rise handlines to operate under low pressure. The two main variables that have been scrutinized are the diameter of the hose and the nozzle pressure. Both of these are key to reducing pressure requirements because of the low standpipe residual pressures. Firefighters have been able to overcome this problem as much as possible with 2½" hose and 50 psi nozzles. For maximum flow based on a large volume of fire, however, there are issues that must still be addressed.

This section emphasizes being proactive with all things necessary for a high-rise pump operation. In regard to high-rise firefighting, if there is a confirmed fire, the water supply portion of the pump operation needs to be set up for a worst-case scenario, which would be a high flow water demand. On a normal fire, not a high-rise fire, the fireground is a little more forgiving when it comes to not maintaining the proper water supply volume. When fighting a fire in a high-rise building, you have one chance to get it right simply because the fire department connections are the only way to get water into the standpipe. If a high-rise operation is not producing the required flow because it was set up for a standard high-rise operation, it may be necessary to shut down some units to obtain more water.

To better understand the techniques for pumping large volumes of water to a high-rise system, review the building codes for fire protection systems and how they work. It may also be helpful to understand the standard tandem pump operation. *NFPA 14, Standard for Installation of Standpipe and Hose Systems,*[1] outlines pressure limitations for building fire protection systems that directly affect how the fire department delivers water within the structure. This standard is updated regularly, but according to the 2019 edition, new systems should be hydrostatically tested at 200 psi or 50 psi in excess of the system pressure, whichever is greater. The maximum allowed system pressure is no higher than 400 psi, with exceptions allowed as detailed by the current standard.

There are two basic sets of standards that have been established by the NFPA that affect water delivery. The first standard was in effect until 1993, when the second standard, which is still in place, went into effect. Structures with the pre-1993 standards were not required to upgrade to the newer standards.

Pre-1993 standards

There is a 500 gpm flow requirement from the hydraulically furthest standpipe from the building pump using two 2½" standpipe outlets with a residual pressure of 65 psi. Where the residual pressure at the standpipe outlet exceeds 100 psi, a pressure regulating device with a limit of 100 psi is required. When the system pressure exceeds 175 psi, a pressure regulating device is required with a limit of 175 psi that works under static and residual pressures.

Post-1993 standards

There is a 500 gpm flow requirement from the hydraulically furthest standpipe from the building pump using two 2½" standpipe outlets with a residual pressure of 100 psi. When the system pressure exceeds 175 psi, a pressure regulating device is required with a limit of 175 psi that works under static and residual pressures.

Authority having jurisdiction

The authority having jurisdiction cam approve standards for their cities that supersede the NFPA standards. For example, Clark County, Nevada (Las Vegas) requires a standpipe residual pressure of 125 psi.

High-Rise System with a Building Pump

Let's review the basic tandem pump operation and why it is needed. Should a combination sprinkler and standpipe system (with a building fire pump) in a high-rise building be overtaxed by fire or fail for whatever reason, fire departments must have the ability to take over the pumping of the system with fire department engines to adequately supply sprinklers, handlines, or both.[2]

Supplying an adequate amount of water to the fire floor in a high-rise building by fire department engines can be divided into two phases. The first phase involves all operations that are required to move the water from the city water mains to the engine or engines that will be supplying the system. The second phase involves all operations required to get the water from the fire department engine or engines into the fire department connection (FDC) of the fire building.

If the fire department needs to pump the system in a high-rise building, it will have to match the building pump pressure and possibly exceed it. In high-rise building systems, one-way check valves are located where the FDC enters into the system and also after the building pump on the discharge pipe. Both are on the discharge side of the building pump. Because of these one-way check valves, if the pump is running but not producing adequate pressure or flow volume, the only way the fire department can enter into the system to take over is to exceed the pressure at the check valve. Either the building system pressure or the fire department engine does the job, but as a rule, they do not work together. If the building pump is not running, there will still be system pressure up against the check valve that will have to be overcome by the engine companies.

> The fire department needs to pump the system pressure no matter what floor the fire is on.

If the building pump should fail to provide the required pressure for whatever reason, the fire department engine will still have to pump the system as though the fire problem is on the roof (i.e., the highest pressure). The reason for this is simple. A building fire pump does not have the capability of controlling its engine pressure over the wide pressure range that would be needed to provide the proper pressure to all the floors in a multistory building. Therefore, the building pump is designed to pump to the highest point in the building. Because of the potential of high pump pressures in the system, there needs to be a way to reduce the pressure at the lower floors to provide the proper pressure to the handlines. This is done through different types of pressure regulating devices. These devices are designed to reduce the building pressure to the proper pressure on the floor where they are in service.

Tandem Pump Operations

A tandem pump operation is nothing more than a relay pump operation, which is one engine pumping water to another. The reason for a relay is to move water from a water supply to an engine through hose when the water supply does not have enough pressure to do so on its own. These operations are usually conducted over long distances. In a tandem pump operation, the purpose is to develop more pressure than a single engine is capable of producing, or at least producing efficiently. These engines are placed close together, with the goal of using 50' sections of hose between them.[3]

Procedures for High-Rise Pump Operations

The following outlines the procedures to follow when given the order to pump the system of a high-rise building:[4]

1. Establish a water supply.
2. Spot the engine or engines that will be supplying the FDC in a position or positions to accommodate the following objectives:
   a. Use a minimum of two discharge lines to make the connection from the discharges of the engine to the FDC.
   b. Spot the engine with the pump panel away from the fire building to help protect the engineer from falling debris.
   c. Connect the discharge lines supplying the FDC and other engines in the operation to the discharges opposite the pump panel, allowing the engineer to stay clear of the lines in the event of a hose failure. If the pump panel is top mounted, there still needs to be a safe side clear of hose for the operator and anyone else who needs to have access to the unit.
3. After all lines are connected, determine the system pressure and set the engine pressure of the engine supplying the system at 50 psi below the system pressure. Setting the FD engine pump discharge pressure 50 psi below the building system pressure will allow the building pump to stay running and do the job as long as it can. A pressure drop in the system of more than 50 psi is a good indication of a substantial amount of water flowing, which in turn is a good indication of a drop in residual pressure that could affect the flow to the handlines. It might also be an indication of a pump failure. It is at this point the FD engines should take over the building system.

The other train of thought is to have the FD pumping apparatus match the system pressure right away to make a more failsafe system. You cannot go wrong with this tactic. In my personal experience with high-rise fires, the overwhelming majority have been small, only requiring one line to extinguish. The building pump can handle this operation. With this being said, why take over the pump when it is working? This is a fire department preference that needs to be reflected in departmental operating guidelines.

Here is a good example of the reasoning behind using pressures 50 psi below system pressure, validated by the Riviera Hotel flow tests (table 6–1).

Table 6–1. Flow test data for high-rise pump operations

Total flow—992 gpm			
Four handlines	1⅛" tip	44 psi average NP	248 gpm
Average standpipe residual pressure—48 psi			
Building static system pressure—185 psi			
Building pump residual pressure—165 psi			

4. Once the fire department engine is set up and pumping its required engine pressure, it should be determined whether or not the engine is flowing water into the system. Whenever the engine starts flowing water, whether it is initially after the engine goes into operation or at any time during the incident, or if the firefighters inside request more pressure, immediately *pump the system pressure.* After the initial hookup, command should be notified whether or not the engine is flowing water and also at any time during the incident if water should start flowing.
5. One method for determining whether or not water is flowing from the fire department engine into the FDC is to partially gate down one of the discharges being used. If the lines are static (water not flowing), the pressure on the discharge gauge will not change. On the other hand, if there is water flowing from the engine into the FDC, the pressure on the discharge gauge will change when the discharge is gated down. It should be noted that a minimum flow from the engine may not show as a residual pressure with the shutdown of one discharge. When conducting this check, therefore, all discharges should be gated down one at a time to determine if the engine is flowing into the system.

If the engine pumping the FDC has a pressure governor, check for a drop in engine speed since the governor will stay at the set pressure in the pressure mode. Better yet, start the operation in the engine speed mode until you reach a point where you no longer have to do the check, or a pumper flow is confirmed. Then the governor should be switched to the pressure mode.

6. If there is an actual fire, additional engines can be requested to duplicate the original pump operation, preferably using a separate FDC bank if available. This will provide reliability for supplying the system as well as additional water flow volume if needed. The clappers in the inlets of the FDC that are not being used can leak under pressure, which will cause pressure buildup in the inlet if the cap is left on. This can be a problem for firefighters who are trying to remove additional caps to add more lines to the FDC. For this reason, all caps should be removed from the FDC before the initial pump operation is set up.

   An alternative method for providing reliability to the pump operation is to add an extra engine in the primary pump operation. If a single engine is all that is needed to create the system pressure, add a second engine in tandem for backup. If a tandem operation is initially set up, add an additional engine, again for backup.

7. Pumps need to circulate water to keep from overheating. If the engine is throttled up but is not flowing water into the FDC, open an unused discharge and its corresponding drain valve. This should flow enough water to keep the pump cool.

The Required Hose

The high system pressures created by taller buildings can exceed the standard hose pressure ratings. Most of the 2½" and 3" hoselines in service today have an annual service test pressure of 400 psi. The NFPA rule has been that the maximum allowed operating pressure for hose should be 10% less than the annual service test pressure. As a result, 400 psi test pressure hose can be operated at 360 psi. (Standards for hose and hose testing can be found in NFPA 1961 and NFPA 1962.[5]) Always consult with the manufacturer to determine what is safe for their products. For system pressures that exceed standard hose ratings, special high-pressure hose should be used. It is not uncommon to have system pressures over 350 psi. Several hose manufacturers make high-pressure hose capable of supporting operating pressures as high as 600 psi.[6]

It is essential that the hose used for high-rise pump operations is reserved for that use alone. The reason it should not be used for normal operations is that hose can get slight cuts, abrasions, and burns that may not burst under normal operating pressures. This may not hold true under the high pressures used in high-rise pump operations, however.

Intake Valves and Pressure Dump Valves

If the incoming pressure to the intake pressure dump valve exceeds its maximum setting, a significant amount of pressure above that point will be expelled through the pressure dump valve instead of contributing to the required engine pressure. Furthermore, there is a possibility that the maximum pressure limitations to the valve body will be exceeded, creating an unsafe situation. Consult with the manufacturer of your valve for pressure limitations. This does not mean they cannot be used under any circumstances. For example, if units with 300 psi intake valve and pressure dump valve ratings are involved with a 500 psi operation, it does not mean that the valves will have the full 500 psi during operation. The pressure needs to be divided between the engines so that the pressures do not exceed the valve capabilities.[7]

Example

A tandem pump operation is being set up to provide 500 psi system pressure using three engines. These engines are equipped with 300 psi rated intake valves both for the valve body and the pressure dump valve attached to it. The first engine pumps 150 psi to the second engine. The second engine pumps 300 psi to the third engine, and the third engine pumps the required 500 psi to the system. At no time does the pressure at any of the intake valves exceed the maximum allowed 300 psi. It is important to max out the adjustment of the pressure dump valve to keep from losing discharge pressure. Some pressure dump valves are capable of being turned off.

Most engines also have built-in pressure dump valves on the steamer intake pipe that must be dealt with if overpressurized (fig. 6–1).[8] The easiest way to do this is to cap off the dump port with a 2½" cap. The dump port will have a label on it that says DO NOT CAP. If the port is capped, the dump valve will not work. That is exactly why we are capping it. Capping it will not damage the dump valve. It will, however, eliminate water hammer safety at that point because it will not activate. Thus, capping off the dump port should only be done if there is no other way to create the required pump discharge pressure. In fact, some dump valves are designed to allow them to be turned off if needed (fig. 6–2).[9]

Figure 6–1. Intake valves have maximum allowed operating pressures.

Figure 6–2. A 6" steamer intake pressure dump valve

Fire Protection System Pressure Chart

A data chart can be developed to show all building system pressures for high-rise structures within a department's response jurisdiction (table 6–2). The addition of hydrostatic pressure data for each building results in a valuable source of information that should be readily available to all engine companies that respond to high-rise districts.[10]

Table 6–2. Sample building fire protection system pressures

Building	System Pressure	Starting Engine Pressure	Hydrostatic Pressure
Allure Condos	330 psi	280 psi	380 psi
Atrium Building	135 psi	85 psi	200 psi

Tandem Pumping for Large Flows

The following must be considered for large flows in a high-rise water delivery system from the engine to the nozzle:

1. The water supply
2. The capacity of the apparatus in the evolution
3. The friction loss in the tandem pumping discharge lines
4. The correct pump discharge pressure for the engine pumping into the FDC

Water supply

The water supply problem is the easiest issue to overcome. The concerns are the same as those of a ground floor level structure fire, with the difference being that all of the water will be going to a one engine operation. Know your equipment and your areas and be able to use a supplemental water supply with additional supply lines (fig. 6–3). Most high-rise buildings are within an urban setting and will have the availability of hydrants. Moving the required flow may involve more than one hydrant, and in fact could actually create the need for a relay pump operation. In such situations, being proactive is vital. If the first-in company does a size up and thinks the fire has developed into a large flow operation or may have the potential to grow into one, establishing a large flow water supply should be considered. This is obviously the first-in officer's choice based on the information at hand. I would set up for at least a 2,000 gpm water supply. Remember, this is a goal to reach. It does not mean you have to use the entire 2,000 gpm, or that you will even be able to achieve it. In most cases using large diameter hose and two hydrants will accomplish that goal. Finally, consider the benefits of a high-rise operational guideline that calls for setting up a large flow water supply as standard practice. If you have the units and the means to get the water, you will be prepared to react quickly if large volumes of water are needed. If they are not necessary, you will have taken advantage of a great training opportunity.

Figure 6–3. Large diameter hose can be used on systems up to 185 psi.

Pump capacity of the engine connected to the supply lines from the hydrants

In a tandem pump operation, the required flow will be based on what the first engine in the tandem operation is capable of delivering, since all supply lines in the evolution are connected to the first engine in the tandem operation. The first engine moves water over to the additional engines involved with the tandem operation, which are used to boost the discharge pressure to match the system pressure of the building (fig. 6–4).

City fire departments generally use engines with pump capacities of at least 1,000 gpm or more. For example, consider a worst-case scenario with a 1,000 gpm engine. Remember we are talking about the first engine in this tandem setup, which is connected directly to the water supply through supply lines. It is a proven fact that a centrifugal pump, which is what we use in the fire service, takes advantage of the incoming water under pressure and can actually pump more than it is rated for. There is no reason why a 1,000 gpm engine cannot flow at least 1,500 gpm to 1,800 gpm if it has the water supply to do so (fig. 6–5).

Another factor to consider is keeping the net pump discharge pressure on all the units in the operation no higher than 150 psi if possible. Remember, pumps are rated to flow their capacity at 150 psi net pump discharge pressure. As the net pressure increases past 150 psi, the capacity of the pump actually decreases.

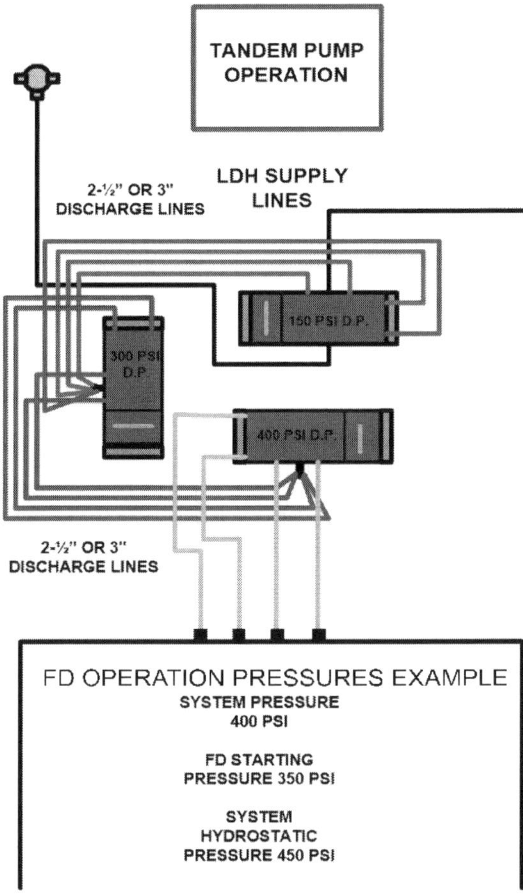

Figure 6–4. Tandem pump operation

This pump discharge pressure should not be too difficult to maintain, but it may help to be aware of this when determining the pump discharge pressures for the units in the operation.

Friction loss in the tandem pumping discharge lines

Large diameter hose, the best hose for moving water efficiently, usually cannot be used in this situation because high-rise pump operations can require high pressures. In some of the shorter high-rise buildings, however, large diameter hose might work. There are two types of large diameter hose in regard to pressure ratings. The most common is supply hose, which is rated at 185 psi. Large diameter

Figure 6–5. The Stratosphere Hotel in Las Vegas has a maximum operating pressure of 600 psi and is pumped by three 1,500 single-stage engines in tandem.

is rated at 275 psi. If the buildings have system pressures that fall in line with this, you could use large diameter hoselines.

Remember we mentioned that in a tandem pump operation, one engine essentially receives the entire water supply. Realistically the amount of water that would be available to a single engine would probably average no more than about 2,000 gpm. This amount of water is capable of being delivered through multiple 2½" or 3" lines with minimal loss (fig. 6–6).

Example. Consider a scenario in which there are three discharges from engine to engine in a tandem pump operation flowing 1,500 gpm. Because a tandem pump operation requires the apparatus to be located in near proximity for safety reasons, you can usually use 50' discharge lines to go from one pump to the other. If you divide 1,500 gpm by three discharge lines, you get 500 gpm per line. The friction loss in 50' of 2½" hose is 25 psi. This pressure loss is not too high and can be overcome with engine pressure. I personally like 3" hose for this type of operation because it has even better friction loss characteristics: 1,500 gpm divided by three 3" discharge lines results in an 8 psi friction loss. Even if your department does not use 3" hose, keeping four sections of 3" hose on the apparatus for high-rise pump operations would be a reasonable practice.

Figure 6–6. Tandem pumping with 2½" high-pressure hose

Another thing to consider is purchasing a siamese intake appliance that has multiple gated 2½" inlets (fig. 6–7). This appliance could be attached to the 6" intake on your apparatus. In fact, you could also put one of these on a front suction, which would really give you plenty of lines. With a little creative thinking, this could be set up to receive six to eight lines. This would reduce the friction loss in the discharge lines to practically nothing.

Raising the required pump discharge pressure to overcome friction loss in the discharge lines for a large flow operation

So far the only things that have changed for putting together a high-flow tandem pump operation are the water supply and the multiple discharge lines. What about the required pump discharge pressure? We know that we are supposed to pump the system pressure, but that is based on a 500 gpm flow. For a fire that only requires a small flow such as one or two lines, matching the system pressure on the engine's discharge gauge will more than likely be sufficient. For a large flow requirement, you will still need to match the system pressure. In addition, the engine pumping the FDC must increase the discharge pressure above the system pressure requirement in order to overcome friction loss in the discharge hose going into the FDC. If this is not done, as more handlines continue to open up in the high-rise, the system pressure will drop. The pump operators need to maintain whatever pump discharge pressure they are using. By doing this the system will for the most part maintain its pressure while allowing additional

Figure 6–7. Task Force Tip's gated siamese manifold allows up to four lines to be brought in to the 6" intake via a TFT intake valve.[14]

handlines to be connected to additional outlets. The only thing that will stop the addition of handlines is running out of water from the water supply or maxing out the throttle on the engines.

At the beginning of this section, we talked about all systems being hydrostatically tested. As a refresher, any system up to 150 psi is hydrostatically tested at 200 psi. Any system pressure higher than 150 psi is tested at 50 psi higher than the system pressure. What this means is that when the system was first installed, it was able to withstand the above-mentioned pressures. As you know, we are dealing with systems that may be decades old. It is reasonable to wonder whether or not we should push our luck by increasing the pressures beyond system pressures, given the age of some systems. Considered from another viewpoint, however, if we need more water and we do not have it, we will lose the battle. So should we chance using greater pressures? If you choose not to risk it, the fire attack will have to be carried out with inferior flows. You may or may not have a sufficient water supply.

The decision to operate at the higher test pressure levels must be made on the fireground by the incident commander. In researching this particular issue, I have worked with many fire protection engineers and fire protection system contractors. The majority have said that they would take the extra risk because it would probably withstand the pressure. If there was a failure, in their opinion it would most likely be a leaking joint connection and not a catastrophic failure in the pipe. Again, this is a decision for the incident commander. If the incident commander chooses to do so, I would recommend increasing the pressure between 15 and 25 psi to see if that satisfies the water demands. As mentioned previously, multiple lines between the engines and the FDC will keep the friction loss to a minimum. A minimum of four lines should be established if possible.

Some of the increased pressure will not even make it to the nozzles in use because it will be eaten up in friction loss in the lines going to the FDC. The friction loss occurs because an increase in pump discharge pressure through the discharge lines to the FDC means that the flow will be increased. As the flow is increased, the friction loss in the FDC hose and the handlines will also increase, along with friction in the plumbing within the fire protection system itself. Therefore, even an increase of 25 psi pump discharge pressure will not mean an increase of 25 psi to the nozzle. It is possible that one-half or less of the 25 psi pressure increase will reach the nozzle.

A majority of the apparatus in service today are equipped with electronic pressure governors, which when left in the pressure mode will automatically maintain the set pump discharge pressure. The problem with using it in the pressure mode when pumping engine to engine is that the units have the potential to react to each other, creating too much fluctuation in engine speeds, or "rpm hunting." If the governor is set in the engine speed mode, or rpm mode, this will not happen, but the operator will have to monitor the pressure and react accordingly. You might want to try keeping the engine connected to the FDC in the pressure mode. This would allow the system pressure to be maintained and might not create the rpm hunting, since the other pumps will be in the engine speed mode. I recommend doing drills to become familiar with what will happen in both modes and use whichever mode works best for you.

Duplication of the pumping operation

Another alternative for a big water pumping evolution is to simply duplicate the first pumping operation. Of course, more apparatus will be needed as well as fire department connections on the FDC. The key to using a duplicate pumping operation is to use two separate FDCs for maximum results. Not all structures have this. If you duplicate an operation using one FDC connection with four inlets or less, each evolution will only be able to use two. This means that the friction loss from the engine to the FDC might be too high, which could greatly hinder the flow.

The following diagram of high-flow tandem pump operations (fig. 6–8) shows three different types of tandem pump operations that can be used to achieve high flows. They are the following: pumping at system pressure, pumping 20 psi higher than system pressure, and something of a duplication of operations. Two separate engines, 5 and 6, are pumping into the FDC with two separate tandem pump operations. Engine 2 at the top right is supplying more water to engine 3, which

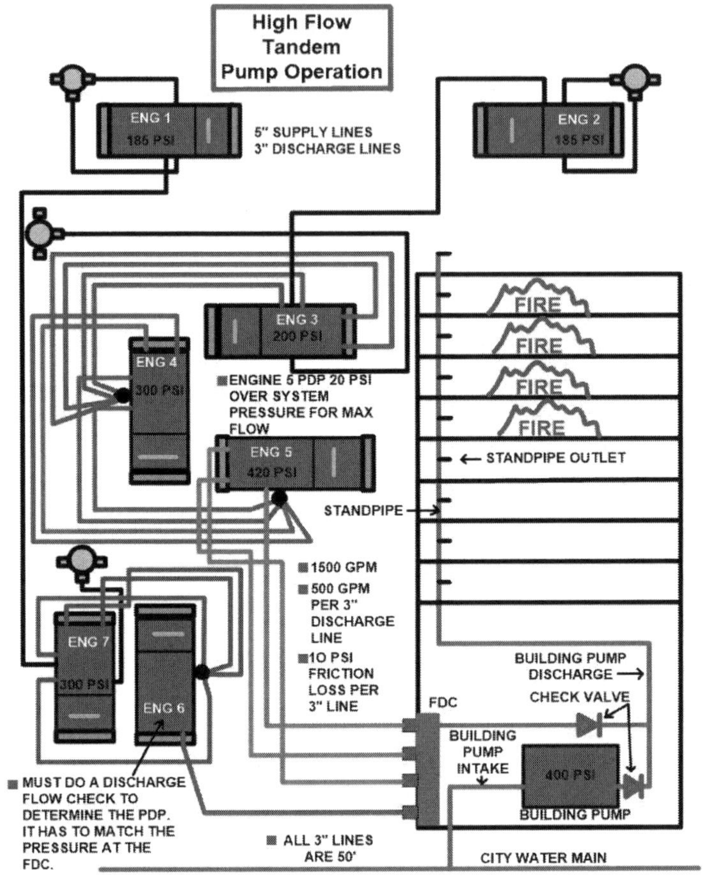

Figure 6–8. High flow tandem pumping operations

is the first engine in the existing tandem pumping operation with engines 4 and 5. There are three engines involved with the second tandem pumping operation, engines 1, 7, and 6. Engines 1 and 7 have hydrants. As you can see, there are many variables that can be used.

The Riviera Hotel High-Rise Flow Tests

A series of flow tests were conducted June 9, 2015 at the Riviera Hotel in Las Vegas, Nevada, with the Clark County Fire Department, Las Vegas Fire and Rescue, Fire

Flow Technology, and Desert Fire Protection, which is a major fire protection system contractor in Las Vegas. These flow tests were conducted to determine the best options for developing a maximum flow scenario in a high-rise structure operation. Part of these tests involved using the system pump only, while the other part involved a tandem pump operation with two engine companies for discharging water into the building.

Guidelines were set up for conducting these tests to ensure that the operation would be as safe and as realistic as possible. All system pressures were strictly adhered to. All flows were delivered through 2½" handlines with 1⅛" tips, with the exception of one that used a 1¼" tip. It was decided to only count the flow delivered by a handline if the produced stream was considered adequate for firefighting. Nozzle pressures were kept at 40 psi and above.

The Riviera Hotel flow tests: First set

The first set of tests was conducted without fire department intervention in the system. The building pump was basically allowed to deliver the flow. There were three goals for this portion of the testing.

Test one. The first was to see how much water one standpipe outlet could produce. The first part of this flow test involved running two 2½" handlines off one standpipe outlet on the top floor (table 6–3). These lines were connected to the outlet via a 2½" × 2½" gated wye.

Test two. The second part of this series of flow tests involved a maximum flow from a single outlet with one 2½" handline (table 6–4).

Test three. The third part of this series of flow tests involved a maximum flow from multiple outlets and three standpipes with 2½" handlines (table 6–5).

Riviera Hotel flow tests: Second set

The second set of flow tests was conducted with fire department intervention into the system. Two single-stage 1,500 gpm engine companies were used in a tandem pump operation pumping into four inlets on the FDC. The goal for this test was to see how much water could be delivered out of the standpipe system

Table 6–3. Two 2½" handlines 150' in length with 1⅛" tips connected to a gated wye, which was connected to the standpipe outlet.

Total flow—476 gpm			
Two 2½" handlines	1⅛" tip	40 psi nozzle pressure	238 gpm
Standpipe residual pressure—65 psi			
Building pump residual pressure—170 psi			
Building static system pressure—185 psi			

Table 6–4. Flow test for maximum flow from a single outlet and handline

Single 2½" handline, 150' in length			
1⅛" tip	265 gpm	50 psi NP	Standpipe residual pressure—80 psi
1¼" tip	345 gpm	56 psi NP	Standpipe residual pressure—74 psi

Table 6–5. Flow test for maximum flow from multiple outlets and handlines

Total flow—992 gpm			
Four handlines	1⅛" tip	44 psi average NP	248 gpm
Average standpipe residual pressure—48 psi			
Building pump residual pressure—165 psi			

in workable firefighting streams while keeping within the pressure parameters of the system. The system is equipped with a 1,000 gpm pump with a *churn pressure* of 180 psi, which is based on NFPA standards that this system followed. (Churn pressure is the pressure in a standpipe system when it is at system pressure but not flowing.) The system also has a hydrostatic pressure of 230 psi, which is 50 psi above churn pressure.[11]

Two 5" supply lines were brought in from one hydrant. Four 2½" discharge lines 50' long were stretched between the two engines and from the second engine to the FDC. A special intake valve designed by Task Force Tips, a company that specializes in water flow equipment, was used for this test. The intake valve made it possible to run all four discharge lines from the water source engine to the engine pumping into the FDC. One of the problems noted in preparing for this test was that most engines are not designed to allow multiple 2½" supply lines to be taken in on the officers' side of the unit. Remember the lines used in a tandem operation can be quite high in pressure, so for safety reasons they should not be brought into the pump panel side. The specially designed intake valve eliminated this problem. In fact, an engine with a front or rear intake could theoretically supply eight lines. It may seem like overkill, but if the pump operation happens to be set up in the debris field, there is a good chance that those lines could be cut by falling debris. Running extra lines between the units would allow for the best chance of an uninterrupted water supply if a line was ruptured.

Riviera Hotel flow tests, second set: Test one. The first portion of these tests involved the tandem pump operation throttling up to the 180 psi system pressure (table 6–6). The pump discharge pressure was maintained throughout the evolution. In other words, after initially throttling up to the required pressure, if that pressure dropped because of an increasing amount of flow from the system,

Table 6–6. A maximum flow through multiple handlines from three standpipes using a two engine tandem pump operation. The second engine in the operation matched the building system pressure of 180 psi.

Total flow—952 gpm			
Four handlines	1⅛" tip	40 psi average NP	238 gpm
Average standpipe residual pressure—60 psi			
Pressure at the inlet of the FDC—180 psi			
PDP of the second engine—180 psi PDP			

Table 6–7. A maximum flow through multiple handlines and a portable master stream from three standpipes and a two-engine tandem pump operation. The second engine in the operation had a PDP of 200 psi.

Total flow—1,315 gpm			
Four handlines	1⅛" tip	40 psi average NP	238 gpm
Master stream	1⅜" tip	42 psi NP	363 gpm
Average standpipe residual pressure—55 psi			
Pressure at the inlet of the FDC—175 psi			
PDP of the second engine—200 psi			

pump operators would actually maintain the required pressure by throttling up, thus allowing for a maximum flow. An engine with an electric governor would do this automatically in the pressure mode. The engines used in this test did not have pressure governors.

Riviera Hotel flow tests, second set: Test two. The second portion of the series again involved a tandem pump operation, but this time the required pressure was increased from 180 to 200 psi (table 6–7). Although the system was pressurized past the system pressure, the hydrostatic pressure limitations were not exceeded.

Comments on the flow tests

There were several interesting points observed from the flow tests. The results of the building pump tests surpassed expectations. The flow test with the two 2½" lines from one outlet using a gated wye that flowed 476 gpm far exceeded expectations (fig. 6–9). The NFPA standard calls for a total of 500 gpm being flowed from two outlets.[12] A flow rate of 476 gpm was achieved from just one outlet.[13]

The single 2½" handline was also impressive. As explained previously, the 65 psi standpipe residual pressure was chosen because it would supply 100' of 2½" hose with a 1⅛" smoothbore tip at 50 psi nozzle pressure flowing 265 gpm

Figure 6–9. Two 2½" handlines connected to one outlet with a gated wye

(fig. 6–10). I would venture to say that most high-rise handlines are set up for this exact scenario, and the real truth is that they are robbing themselves of a significant amount of water if the line was intended to be a high-flow line (265 gpm or more). Again, the test achieved 345 gpm with a 150' handline using a 1¼" tip.

The final flow test for maximum flow with multiple 2½" handlines is probably the most impressive test conducted in the entire series. The test achieved 992 gpm from a building pump that was rated to flow 500 gpm with a 65 psi standpipe residual pressure, nearly doubling the rated amount. The thing to remember is that multiple standpipes increase the flow from the pump.

This system is the worst-case scenario for an NFPA fire protection system. It is a pre-1993 building system rated to flow 500 gpm at 65 psi residual pressure. The flow tests were conducted with the most remote part of the system, the roof.

The results from the second portion of the flow tests conducted from a tandem pump operation were also eye-opening. The flows were 952 gpm at 180 psi and 1,315 gpm at 200 psi. The original plan called for conducting an additional flow test with a tandem pump operation 50 psi over system pressure. This test was not able to be carried out because the hydrant was unable to supply additional water. Had this been a real fire, however, the need for more water could have been met by laying an additional supply line from a second hydrant and then connecting it to either engine in tandem.

One interesting consideration to note about fire protection systems and the rated capacities of the in-house pumps is that while these pumps are rated to get a specific flow volume, requirements for delivering the flow do not always allow for working fire streams to match the rating. The building pump for this system was rated at 1,000 gpm. When the system was tested, the water was discharged through three 1¾" smoothbore tips with a nozzle pressure of 15 psi. The 15 psi nozzle pressure will not produce a firefighting stream. It is almost like having an open butt discharge flowing the water. Again, the test used actual 2½" handlines

Figure 6–10. All standpipe outlet residual pressures were recorded with an in-line gauge.

with 1⅛" tips and required a workable stream for the flow from the specific line to count in the tests. It is one thing to produce a large amount of water, but it will not matter if you cannot use it.

It is important to note that the information taken from these flow tests is unique to this building only in regard to the exact numbers. The results should offer a reasonable guide concerning what to expect from a high-rise building.

Remember, in a high-rise fire we are already starting behind the eight ball in regard to water delivery. The techniques presented in this section have been proven on the fireground and are relatively safe to undertake, but do not expect miracles in regard to several thousands of gallons of water per minute. I recommend that departments with high-rise buildings in their jurisdiction formulate a plan for flowing big water in a high-rise structure. Start with your fire prevention bureau. Contractors of fire protection systems are also excellent resources simply because they design and build the systems. There is a great deal involved with an operation like this, and it is possible that you may only respond to one high-rise fire in your career, if that. As such, it is extremely important to conduct realistic training on a regular basis.

7

WATER MANAGEMENT OFFICER

Introduction

Battling a working fire requires successfully coordinating multiple tasks. One of the most important parts of the overall strategy that has the potential for ensuring the success of an operation is being able to obtain adequate water delivery. In this discussion, the term *water delivery* is deliberately used instead of *water supply. Water supply* implies that we are simply interested in finding the water, but that is only half the battle.[1]

Producing adequate firefighting streams at large fires can be a complex operation involving the efforts of many people, numerous pieces of equipment, and hundreds or sometimes even thousands of feet of fire hose. A plan of operation is needed to effectively coordinate these fireground tasks. In fact, a major water delivery operation can be worthy of its own command structure.

For effective fireground operations, one person—the water management officer (WMO)—should be in charge of the overall fireground water delivery system. This will avoid freelancing efforts from others on the fireground and will allow a structured water delivery operation. The duties of the WMO are simple and yet require knowledge of several areas of water movement on the fireground. The sole responsibility of the WMO is to manage water delivery on the fireground at its present state, as well as to prepare to respond to any possible escalation of water delivery demands. The WMO must be aware of water supply and water discharge.

Water Supply

The WMO must first ensure that there is an adequate water supply to the fireground for the present stage of the fire, as well as prepare for a possible escalation. There are several factors that the WMO should be aware of to assist in maintaining the water supply.[2]

The static pressure of the hydrant system being used should be noted, but this information is only half of what is necessary to determine the strength of the hydrant system. Obviously, the higher static pressure will have a better chance of being able to move water farther along the supply hose. The other half of what needs to be ascertained is the volume of water available from the water main. A high static pressure on a low-volume water main will usually not produce big water volumes. A good example of this is a hydrant system that I have used that had 150 psi static pressure but was supplied by a 3" water main. As soon as this hydrant was opened, the residual pressure dropped to almost nothing, flowing only 400 gpm. On the other hand, I have seen a hydrant system with 65 psi static pressure that still had 45 psi residual pressure after flowing 1,500 gpm.

The residual pressures of the pumpers working at the fire should also be noted. A low residual pressure is an indication of low incoming water to that unit (fig. 7–1). When small diameter hose (2½" and 3") is used for the supply line, a low residual pressure that needs immediate attention is about 20 psi. On the other hand, if large diameter hose is being used, especially 5", the residual pressure can usually be brought down to 0 psi in a lack of water situation. This is not to say that a low residual pump intake pressure with the use of large diameter hose is acceptable or should not be addressed.

It is important to realize that a low pump intake residual pressure indicates a low water volume at the pump but not necessarily at the hydrant. Often the supply line is the restricting factor in delivering more water to the pump. Large diameter hose offers the best means to deliver maximum flow.

To obtain maximum flow to a single engine company from a hydrant system, multiple supply lines from one or more hydrants alone or in conjunction with a relay pump operation should be considered. If a relay pump operation is being considered, make sure that there is a sufficient number of pumpers and hose to deliver the required volume of water from potentially long hose lays (fig. 7–2).

Occasionally at fires requiring large water flows, the initial responding pumpers will bring in supply lines from the closest hydrants. As the incident and the water flow requirements escalate, there is a good chance that the first-in units could start exhausting their initial supply lines (fig. 7–3). These units are often in a good location to deliver the water in the fire attack and have the ability to produce more fire streams, but they may need more water. Individual relay pump

Figure 7–1. A low intake pressure at the pump panel indicates a low volume at the pump but not always at the hydrant.

Figure 7–2. In this photo you can see under-pumped streams due to a lack of water in the supply lines.

Figure 7–3. The single-engine company operation initially received water from a hydrant and depleted its water supply. The unit was in a good position and had the means to flow more, but it needed more water. Additional supply lines were set up in a relay pump operation to supplement the engine water supply.

operations can be set up to supplement multiple pumpers at the fire using dual pump operations and/or manifold operations (fig. 7–4). It is important to know the number of intakes on each pumper as well as their compatibility to specific hose sizes. This will help determine the type and number of supply line operations that must be established (fig. 7–5).

These types of water supply operations can take several minutes to set up. Do not wait until a water supply problem occurs. If the water supply is dwindling and the fire is not subsiding, initiate the appropriate supply line evolutions to bolster the water supply.

To operate effectively, it is helpful to have access to water supply maps. These maps should include hydrant locations, main sizes and locations, and potential static water sources. The WMO should be aware of specific water supply problems that may be associated with the operation, such as low pressure/low volume hydrants, lack of hydrants, dead-end mains, etc. Finally, the officer must also be aware of all potential mutual aid companies and their equipment that could be pertinent to establishing water supplies, making sure that compatibility between agencies exists.

Figure 7–4. A relay pump operation is supplying two engines with the help of a dual pump operation.

Figure 7–5. A relay pump operation is feeding into a 5" gated manifold, which in turn is supplying two engines.

The WMO must have a thorough understanding of the municipal water system in all areas that are served by the fire department. There should also be a good working relationship with the water department, especially during a major fire when the water requirements of the system may need to be boosted.

Water Discharge

Getting water to the pump is only half the battle. It is equally important to be able to discharge the water onto the fire by means of an adequate fire stream. Let's see what it takes to achieve this goal.[3]

The WMO must be aware of the discharge capabilities of all units involved in the operation. If more water is requested from a pumper already discharging, and the water supply is sufficient, a decision must be made as to whether or not it is capable of moving more water. There are several indicators that will help determine how much more water will be available, if any. Centrifugal pumps have the ability to deliver much more than what they are rated at when supplied by a positive pressure water source such as a hydrant system. In fact, in some cases the capacity can be doubled. Knowing the rated pump capacity as well as the current flow will give you an idea of what might be available from a pumper. As a rule of thumb, as long as the water supply is adequate, expect at least 800 to 1,000 gpm above the rated pump capacity.

It is also important to check the engine speed of all pumpers involved with the operation. During high-flow operations, the engine of the apparatus involved with the operation will often be working at higher speeds. A check of the unit's tachometer will help determine if more water can be produced. If the engine speed is approaching the governed speed of the engine, more water may not be available.

Sometimes the hand throttle on the pump panel will limit the amount of additional water available from a pump. At high-flow operations, the hand throttle on the pump panel can actually be turned out to a point where mechanically it will not turn any farther.

Discharge evolutions, including the hose lays and the nozzles, should be evaluated by the WMO, with corrections being made as needed (fig. 7–6). If the wrong size hose is being used, the friction loss may be too high to deliver the required flow. The wrong size tip on a smoothbore master stream operation not only could interfere with the proper water requirements, but it also could produce an inferior stream.

Fireground Water Management Checklist

Water supply

- Static pressure of the hydrant system[4]
- Residual pressures of the pumpers involved in the operation
- Hose size of supply lines
- Quantity and sizes of intakes on the pumpers involved in the operation
- Water supply maps
- Specific water supply problems
- Possible need for a relay pump operation
- Mutual aid companies and their compatibility to the operation

Water discharge

- Approximate flows from the pumpers involved in the operation
- Pumper discharge capabilities
- Engine speed (rpms) of the pumpers involved in the operation
- Discharge evolutions on the fireground

Figure 7–6. The WMO is responsible for making the streams count.

Conclusion

The end result of any fireground large flow water delivery operation is to provide the required volume of water in the form of effective fire streams to reach and overwhelm the heat energy produced by the fire, achieving a knockdown as quickly as possible. In order to do this efficiently, it is important for the water management officer to thoroughly understand and implement the various aspects of a large flow water delivery operation. The water management officer has a vital role in the command structure of a fire to oversee water supply and discharge, and the importance of this position should not be overlooked.

8

COMPARING 4" AND 5" LARGE DIAMETER HOSE

Introduction

The majority of municipal fire departments are using either 4" or 5" large diameter hose for their water supply hoselines. There are a number of considerations to review when a fire department is choosing the size of large diameter hose to purchase. Some of the considerations involve nonperformance aspects of the hose, such as the relative weight and cost of the hose. These are certainly not the only important considerations, however. A purchase decision based solely on the weight and cost of the hose could limit water delivery capabilities.[1]

Before we get into comparing the performance of 4" and 5" hose, it is worth mentioning some of the comments that have been made by various departments about the size of the hose that they are using:

1. "A 4" hoseline can supply our pumpers with all the water we need."
2. "With all of our hydrants located within 300" of each other, 4" hose is big enough to give us all the water we need."
3. "A 4" hoseline is easier to pick up after it has been used."
4. "If a single 4" line from the hydrant cannot supply the water needed, a relay pump operation can be put into service with the use of a second pumper and a four-way valve."
5. "LDH (4" or 5") is not used because we only have low-volume hydrants, which will not support the big hose."
6. "LDH (4" or 5") is not used because at most of our major fires, responding units may be required to run over supply lines, and that cannot be accomplished with LDH."

7. "LDH (4" or 5") is not needed because our department has many engine companies, so if big water is needed, it can be delivered in a relay operation."
8. "We don't have the luxury of a lot of engine companies to tie up in a relay operation. That's why we use 5" LDH, which gets us all the water we need, and with just hydrant pressure."
9. "We found after evaluating 4" and 5" hose that 5" hose usually could deliver capacity flows straight off the hydrant, while 4" could not."
10. "One reason for choosing 4" hose over 5" hose was because we knew that once the 5" hose was charged, it could not be moved if necessary."

Water Delivery Capability

The most important factor for the fire department to consider when making the decision of which size hose to use should be water delivery capability. Analyzing water delivery capability will require a review of water supply demands, hydrant performance and spacing, water delivery operation tactics, and the number of pumpers available for water delivery operations.[2]

Here are some simple facts about 4" and 5" large diameter hose:[3]

1. There are two types of large diameter hose, *supply hose* and *attack hose*. Supply hose has a maximum operating pressure of 185 psi. Attack hose has a maximum operating pressure of 275 psi.
2. In 100" sections, 5" hose weighs approximately 105 lbs while empty, and 4" hose weighs approximately 80 lbs empty in the majority of cases. There are some manufacturers, however, that offer both 4" and 5" hose in a lightweight version weighing several pounds less in either diameter.
3. Large diameter hose can be used for either supply lines or discharge lines.
4. A 5" hoseline flows twice as much volume as a 4" hoseline.

Before we proceed further with our comparisons, let's analyze statement number four. It is a fact that 5" hose flows twice as much water as 4" hose, at least on paper. In reality this does not always hold true. In order to prove that 5" hose flows twice as much water as 4" hose, you will need to have sufficient water available from the water supply and also the necessary pump capability. Consider an example with a hydrant that has a capacity to flow 1,500 gpm. The testing evolution is a 300' section of 4" supply line from the hydrant to an engine. We throttle up the engine flowing water to reach maximum flow, as indicated by the intake gauge, and we note that the flow is 900 gpm. Next we repeat the evolution with a 5" hose. It will be impossible to double the flow from the 4" hose because the water is not available from the water source. Even if you have 10" hose on the ground, you still can only get what the water supply gives you. If we are drafting from a static water source, and the pump capacity of the unit drafting is 1,500 gpm, then that is all we will get.

Below is an illustration of flow tests that were conducted to match 4" hose against 5" hose (fig. 8–1). Note the first evolution for each size, which is a single supply line from the hydrant to the engine. With the 4" hose, the flow rate was 921 gpm. If we were to double that to show that 5" hose flows twice as much water, it would be 1,842 gpm. As you can see on the 5" hose evolution, we did get more water, but it was not twice as much. We flowed 1,386 gpm. This is still a great improvement, and as far as I am concerned, it is worth using 5" hose instead of 4" hose to achieve this result.

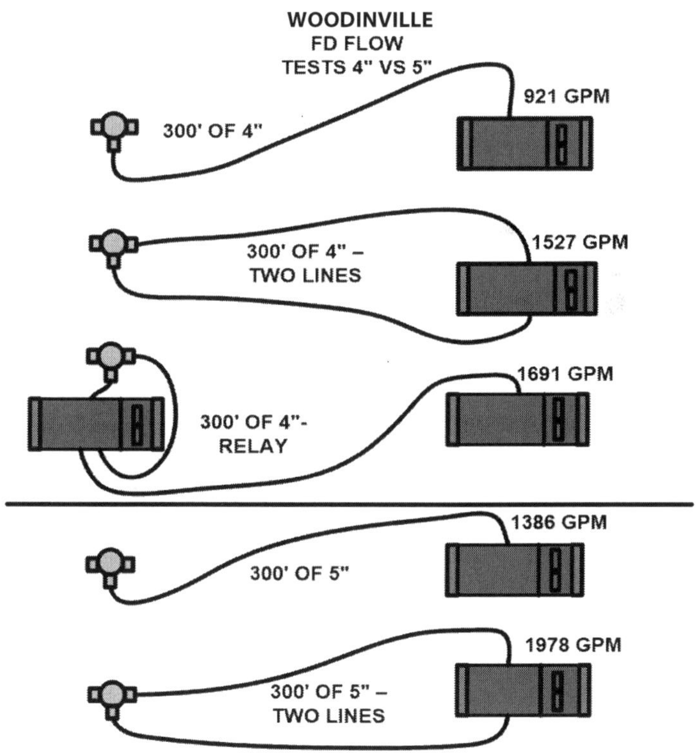

Figure 8–1. Flow tests comparing 4" and 5" lines show that doubling supply lines does not mean doubling the water flow.

Supply Line Evolutions

Before we begin our comparison of 4" hose and 5" hose in specific supply line evolutions, we need to quickly review fire hydrant systems and how they support these evolutions. A fire hydrant operates from water mains filled with water under pressure, which moves the water. As explained in chapter 1, a hydrant system has two basic pressure ratings, static and residual. The residual pressure, which is the pressure while the system is actively flowing, will determine how well the supply hose will be able to deliver the required flow. In addition, there must be a sufficient volume of water available whether the pressures are high or low.[4]

Let's take a look at the most common supply line evolution deployed by fire departments, the house fire scenario in which a single supply line is laid from a hydrant. In most cases the flow will be 500 gpm or less. In this case, the hydrant being used has 60 psi static pressure. After flowing 500 gpm, the residual pressure is down to 40 psi. This measurement represents the pressure that will move the water through the supply hose to the pumper at the fireground. The friction loss in 4" hose flowing 500 gpm is approximately 5 psi. If you divide 40 psi by 5 psi, you come up with 8. This means that this hydrant will flow 500 gpm 800' using 4" supply hose.

Now consider what 5" hose will do with the same hydrant. The friction loss in 5" hose flowing 500 gpm is roughly 2 psi. In reality this would not show up on a discharge gauge, but we will use this number for comparison purposes. Divide 40 psi by 2 psi, and you come up with 20. This means that 5" hose will move 500 gpm with the 40 psi residual hydrant pressure 2,000'. This is slightly more than twice the distance of the 4" hose.

Relay Operations

Is it possible to use a single supply line evolution connected directly to the hydrant with a flow requirement of 500 gpm for a distance of more than 800'? If the possibility exists, and your department already employs relay pumping to boost the pressure/flow in your supply lines, then 4" hose will work for you. On the other hand, if your department wants to rely on hydrant pressure for your supply line evolutions, then the answer would be 5" hose.[5]

To continue with the 500 gpm flow scenario, how far can you move 500 gpm in 4" and 5" large diameter hose set up in a relay pump operation based on supply hose with an operating pressure of 185 psi? A 4" LDH relay will move 500 gpm 3,700', while the 5" relay will go a whopping 9,200'. This is only a two-engine relay, one at the water supply and one at the fire. Based on this example, 4" hose would be sufficient.

Large Flows

So far we have discussed low flow operations. Based only on this information, it would be tough to recommend 5" over 4" hose, especially when it costs more money and is heavier. Some departments may choose 4" hose based on the reasoning that they do not respond to big fires in their districts. This type of thinking will only get you into trouble. Fire departments and firefighters are sworn to protect against all fires, not just the small ones. When the big one hits and you are not prepared because you purchased the wrong equipment or did not use the proper hose evolutions, or both, you will be unprepared and probably unsuccessful in extinguishing the fire.[6]

Now let's step up the flows to see how the 4" and 5" hoses perform. Our goal is to reach 1,000 gpm. Most fire departments that use 4" and 5" large diameter hose lay directly from the hydrant to the fire no matter how long the lay is or what the flow requirement is. There are two misleading assumptions that justify this action. The first is, "We have an excellent hydrant system to support our hose evolutions, allowing us to utilize hydrant pressure alone to meet our demands." A good hydrant system is definitely a plus, but you have to remember it is what comes after the hydrant that causes restriction, mainly hose. For example, consider a flow test on a hydrant without anything connected to it that resulted in a flow of 2,060 gpm. After connecting 300' of 5" hose to it, we were only able to get 1,250 gpm. The reason for this is the restriction in the 5" hose.

The second misleading assumption is, "Laying a large diameter hose supply line is like laying an aboveground water main or like taking the hydrant right to the fire." The large diameter hose is considered a cure-all, and simply hooking it to the hydrant is expected to bring in as much water as is available. This assumption likely dates back to the use of smaller supply lines and the noticeable improvement with larger diameter hose. A comparison between 3" hose and 5" hose reveals that it would take four 3" supply lines to move as much water as a single 5" supply line. It is no wonder firefighters back in those days thought large diameter hose was the best thing to happen to them since motorized apparatus. The true fact is that large diameter hose, as good as it is, will probably not max out the water supply from a decent hydrant by utilizing just hydrant pressure. This especially holds true for 4" hose.

In another example, we can compare 4" and 5" hose on a 600' supply line connected directly to the hydrant. The flow is 1,000 gpm, and the residual pressure in the hydrant after flowing 1,000 gpm is 60 psi. Divide 60 psi residual pressure by 20 psi, which is the friction loss for 4" hose at 1,000 gpm, and you will see that 60 psi will only move 1,000 gpm 300'. Using the same 60 psi residual hydrant pressure and 5 psi for the friction loss in 5" hose at 1,000 gpm, we will be able to move

1,000 gpm a total of 1,200', thus successfully delivering the water in the 600' supply line example. Keeping these numbers in mind, if your department operates from hydrant pressure only, I recommend choosing the 5" hose.[7]

The following flow tests show how building a supply line evolution produces more water (fig. 8–2). We started off with a single line from the hydrant and went to a maximum flow evolution.

As previously mentioned, in order to efficiently fulfill water supply demands, you need to do more than just lay a line from hydrant. Multiengine water supply evolutions need to be set up from numerous water supplies. Relay pump operations are the only way to fulfill the required flows by boosting the pressure to move the water through the supply hose. This holds true for both 4" and 5" hose. This should not be a freelance operation. A water supply officer needs to be in charge of setting up these evolutions, which will involve deploying the lines, spotting apparatus, and choosing the best water sources to get the required flow.

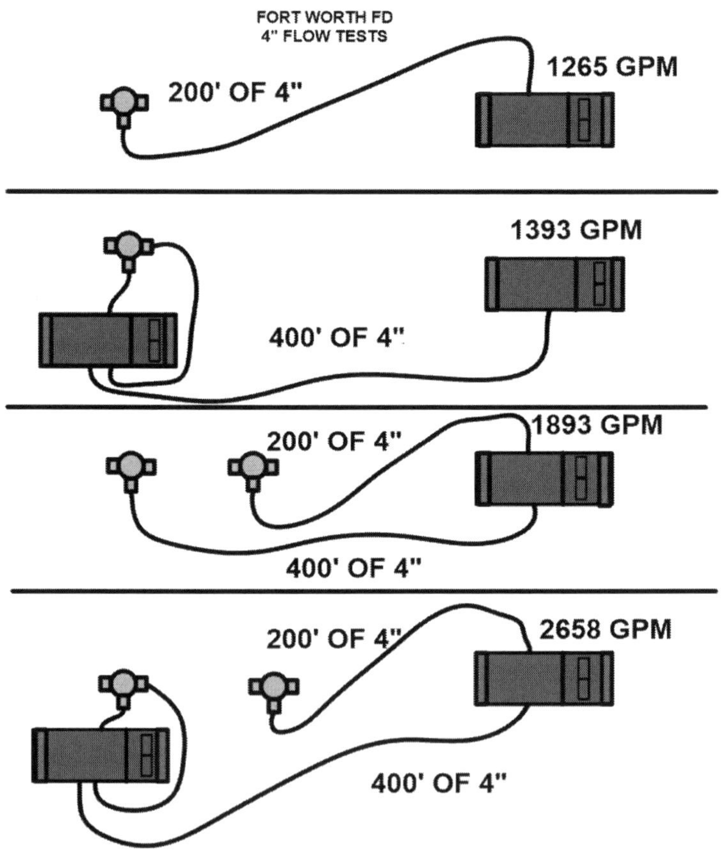

Figure 8–2. Fort Worth FD 4" flow tests

Weight Considerations

Let's briefly examine the weight of 4" and 5" hose. Previously we discussed the difference between the two, which for a 100' section is 105 lbs for 5" hose and 80 lbs for 4" hose. Again, I think this is a secondary decision to be made after all considerations for performance have been evaluated. The weight issue is usually only a problem when the wrong techniques for repacking the hose are used. One suggestion to consider if your choice is 5" hose is to purchase it in 50' sections. The cost will be somewhat higher because of the need for more couplings, but the weight will be reduced by approximately half.[8]

Conclusion

Based on the water delivery capabilities of the supply lines supported by hydrant pressure alone, the 5" hose would be the proper choice. However, if your department supports supply line evolutions with relay pump operations as well as multiple supply lines from the water sources as needed, I think it is fair to say both 4" and 5" hose will be able to move any amount of water that is required. This is still dependent on the availability of sufficient supply from the water source and the presence of the necessary equipment and personnel to set up the proper evolutions. Basically, the difference between 4" and 5" hose would be that the 4" operation would have to work harder than the 5" operation to accomplish the same goal.[9]

II

DISCHARGE WATER DELIVERY

Introduction

Discharge water delivery covers everything from the pump discharge to the stream producing appliance (the nozzle). Besides the actual pumper, two basic types of equipment are involved in producing the firefighting stream—the hose and the nozzle.

The water supply portion of this book goes into great detail to show how to produce the water supply to support any and all firefighting streams. As mentioned in the water supply portion (section I), I will push the limits within the specifications set forth by the manufacturers of the equipment being used to achieve maximum performance. The same guidelines hold true for developing firefighting streams. Some of the information presented about discharge water delivery will not be new to you because these are things we have done for years that continue to provide efficient operations. You will learn about upgraded techniques with current equipment that can increase your operational proficiency. Some of this information may not be in current fire stream management books or may even contradict what the current books and departmental guidelines dictate. The information provided in this book went through extensive research and development that included manufacturers' specifications, flow tests, and evaluation of functionality before being presented to the firefighting community. It is important to note that my studies and the content of the discharge section in this book are based on the mechanics of moving water. This text does not go into detail about hose handling techniques for stationary work.

9

BUILDING A DISCHARGE EVOLUTION

Introduction

The purpose of a discharge evolution is to provide the means to put water on the fire for fire extinguishment through a combination of pumpers, discharge hose, nozzles, and of course, firefighters. Not only does the water have to reach the fire, it has to flow in sufficient quantity to meet and exceed the heat energy produced by the fire to accomplish an efficient extinguishment. This does not happen automatically merely by putting a nozzle on a hoseline or master stream appliance. Certain steps must be taken to create the best possible stream. The first thing you need is the proper amount of water, both in terms of quantity for the duration of the firefighting efforts and in terms of the delivery rate. Pump operators need to support the discharge operation by providing the required pump discharge pressure (PDP) and assisting firefighters in developing the proper discharge evolution. And finally, the proper nozzle must be used, and it must be used correctly.*

In my 35 years of teaching, I have noticed that there is some resistance to pumping high discharge pressures, even though the pressures are within the hose ratings. Several reasons are given for this reluctance, including increased hose stiffness, hose failure, and more difficulty at the nozzle. I even heard one hose manufacturer say that it wastes fuel. It is true that the higher the pressure, the stiffer the hose will get, but I look at that as an advantage. There is less tendency for kinking, especially when going around corners. This can be easily demonstrated by charging a line with the nozzle closed and setting the pressure at what your potential high pressure will be for the line you are pumping. For example, a high-performance 1¾" hose is pumped at 210 psi for a 200' line to get 325 gpm. Set your pressure at 210 psi and conduct a few flexibility tests on it. See

Note: The high-flow handlines discussed in this section are meant to be used as sit-on lines because of the higher nozzles pressures.

how tight you can make a loop in it. Take it through a structure such as your burn tower or maybe even the station (first making sure there are no leaks, of course) to see how the hose performs. I think you will be pleasantly surprised. As far as having difficulty at the nozzle, that notion is exaggerated. The nozzle reaction depends on the flow rate and the pump discharge pressure. The pump discharge pressure must be sufficiently high to get the required flow to the nozzle, but the water does not arrive at the nozzle at the same pressure. The water pressure at the nozzle will be diminished by friction loss in the hose.

There are two basic types of water delivery nozzles in the fire service—the automatic nozzle and the fixed-gallonage nozzle. The latter includes the combination nozzle (both fixed and selectable), the smoothbore nozzle, and most specialty nozzles, such as the Vindicator and piercing nozzles.

Fixed-Gallonage Nozzles

A fixed-gallonage nozzle is designed to flow a specific rate at a specific nozzle pressure. For example, a 1⅛" tip is designed by means of the size of the tip diameter to flow 265 gpm at a nozzle exit pressure of 50 psi. A fixed-gallonage combination nozzle is designed to function the same way, except the internal waterway of the nozzle dictates the flow. For example, a fixed-gallonage nozzle is designed to flow 160 gpm with 50 psi nozzle base pressure. The same holds true for a selectable gallonage nozzle. The only difference is that several flows can be set at the same nozzle base pressure by turning a ring on the nozzle. The last category of fixed-gallonage nozzles includes the specialty nozzles, such as the Vindicator nozzle (fig. 9–1). The Vindicator is a straight stream nozzle only, with built-in stream shapers contained in a large bore tube with a nozzle valve that has a flow range and coinciding nozzle base pressure. All of these nozzles can be underpumped or overpumped, changing the nozzle exit and nozzle base pressures while still producing an adequate stream. Of course, this can only be done to a certain extent before the stream becomes inadequate. As a general rule of thumb, any nozzle pressure under 40 psi is usually not able to provide an effective stream as far as reach is concerned. Even so, you never know until you try. Go out and practice with the equipment. This is the best way to learn.

Smoothbore Nozzle Tips

When it comes to water delivery on the fireground, it is safe to say that the fire service has come a long way from the use of buckets in bucket brigades. Subsequent

Figure 9–1. Fixed-gallonage nozzles include a smoothbore nozzle, a fixed-gallonage combination nozzle, and the Vindicator nozzle.

developments in fire hose and water delivery appliances (nozzles) have greatly improved our capabilities. The smoothbore nozzle has been in service for many years, and we are still using it successfully today (fig. 9–2).

Definition

The smoothbore nozzle is a fixed-gallonage device for producing fire streams. It consists of a nozzle valve and a pipe-like extension that forms the fire stream. It is used in conjunction with handline and master stream operations to deliver water in the form of an effective firefighting straight stream. The flow rating of the smoothbore tip is based on the diameter of the tip as well as the nozzle pressure.

Handline operations using smoothbore tips

Table 9–1 shows the most commonly used handline smoothbore tips, along with their corresponding flow and pressure ratings.

Challenging Conventional Wisdom about Smoothbore Nozzle Use

Rules governing the use of smoothbore nozzles have been in place without change for at least the last 60 years and possibly even longer. I think it is time for a change.

Figure 9–2. The smoothbore nozzle design is still being used today.

Table 9–1. Common handline smoothbore tips and corresponding flow rate and pressure ratings

Tip Size	Flow Rate (gpm)	Nozzle Pressure (psi)
7/8"	161	50
15/16"	185	50
1 1/8"	265	50
1 1/4"	325	50

An examination of the commonly held firefighting guidelines from the International Fire Service Training Association (IFSTA) will demonstrate why this is the case.

The following five rules on smoothbore tip handline operations are from the current fire stream management book by the International Fire Service Training Association:

1. The maximum handline flow is 300–350 gpm. Flows higher than this will create an unmanageable handline.

2. The nozzle pressure for a smoothbore tip on a handline is 50 psi. If need be, the nozzle pressure can be increased to 65 psi. Pressures higher than this will create an unmanageable handline as well as a broken, nonproductive stream.
3. The diameter of the smoothbore tip should be no larger than half the diameter of the hose.
4. When operating a smoothbore handline, the ball valve must be fully open to produce the best quality stream. A partially opened valve will create turbulence in the waterway, breaking up the stream. The ball valve should be fully opened or fully closed.
5. The size of the waterway in the ball valve should be larger than the tip size. If the waterway is smaller than the tip size, it will act as the tip itself, thus creating turbulence as the water enters the larger tip size. This will produce a broken stream.
   (*Note:* This last rule is accurate.)

Table 9–2 shows various nozzle combinations and their capabilities as they relate to handline operations. The chart is divided into sections of flows ranging from 250 gpm to 500 gpm. It shows the nozzle (smooth bore or combination), nozzle pressure, and nozzle reaction.

IFSTA recognizes all combination nozzles listed up to the flow of 350 gpm, as mentioned previously (table 9–3).

Table 9–4 shows smoothbore nozzle pressures and flow combinations that are considered unmanageable due to the nozzle reaction being too high, at least according to IFSTA. Every single one of these tips has an equal or lower nozzle reaction than the highest mentioned combination nozzle reaction that is sanctioned by IFSTA, and yet it is said that they are more difficult to handle and are not recommended.

Nozzle reaction is nozzle reaction. It is the reactive force measured opposite the stream direction (usually denoted in pounds). As long as the combination nozzle is in the straight stream mode, there is no difference.

Nozzle Pressure as It Relates to Stream Performance and Manageability

According to conventional wisdom, an overpressurized stream will lose its pipe-like appearance and begin feathering on the edges, causing a loss in reach

Table 9–2. Nozzle combinations and capabilities for handline operations

Handline Nozzle Reaction Chart			
Flow Rate (gpm)	Tip Size	Nozzle Pressure (psi)	Nozzle Reaction (lbs reactive force)
	15/16"	90	124
248	1"	72	113
	1⅛"	50	99
	Combo	100	125
297	1"	100	157
	1⅛"	64	127
	Combo	100	150
325	1"	120	188
	1⅛"	75	149
	1¼"	50	123
	Combo	100	164
350	1⅛"	87	173
	1¼"	57	140
	Combo	100	177
403	1⅛"	115	229
	1¼"	80	196
	1⅛"	50	148
	Combo	100	204
433	1⅜"	59.5	177
502	1⅜"	80	237
	1½"	56	198
	Combo	100	254

Table 9–3. Combination nozzles and flow rates (350 gpm and lower) considered "manageable"

Flow Rate (gpm)	Nozzle Reaction (lbs reactive force)
250	126
300	152
325	164
350	177

Table 9–4. Smoothbore nozzle pressures and flow combinations considered "unmanageable"

Flow Rate (gpm)	Tip Size	Nozzle Pressure (psi)	Nozzle Reaction (lbs reactive force)
248	15/16"	90	124
252	1"	72	113
297	1"	100	157
326	1⅛"	75	149
431	1⅜"	59	175

and water volume. The first question that comes to mind about nozzle pressures in smoothbore tips, however, is why a handline stream begins breaking up at nozzle pressures higher than the 50 to 65 psi range, while a master stream can be pumped up to 80 psi (again according to conventional wisdom). There should not be any difference.

This can be illustrated with an example. A 400 gpm flow is an accepted stream from a master stream appliance. It is produced from a 1¼" tip at a nozzle pressure of 80 psi. The stream quality is good, and the reach is sufficient. So why can't the same tip and nozzle pressure produce a sufficient stream on a handline?

Flow tests have proven that 80 psi handline nozzle pressures do work just as effectively as they do on master streams. In fact, nozzle pressures reaching 120 psi and higher have proven to be very effective on handlines with the proper nozzle-handling techniques. These higher pressure streams not only give higher flow rates, but they also significantly increase the velocity and the reach of the stream (table 9–5). *Nozzle base pressure*, as referred to in table 9–5, is the pressure created at the base or inlet of the nozzle. It is used to determine nozzle pressures on nozzles that cannot be measured with the pitot gage such as the combination nozzles and the Vindicator nozzle.

Table 9–5. Reach test

Nozzle Type/Size	Nozzle Pressure (NP)/Nozzle Base Pressure (NBP) (psi)	Flow Rate (gpm)	Reach
TFT Automatic	100 NBP	150	90'
15/16"	50 NP	185	84'
1⅛"	40 NP	240	84'
Vindicator Heavy Attack	50 NBP	250	90'

There is a place for the 40 to 65 psi nozzle pressures. If penetration and reach are not factors in the stream application, a low pressure might be all that is needed. This can allow for higher flows for interior attack lines.

Does Low Nozzle Pressure Cause the Hoseline to Kink?

One common issue with low nozzle pressures and smoothbore tips is the fact that the hose kinks more easily. This has been common knowledge in the fire industry for years, especially in high-rise fires. Kirk Allen, president of First Strike Technologies, conducted a series of flow tests that disproved the low nozzle pressure/hose kinking theory in certain cases. The following series of flow tests were designed to support this concept (table 9–6).

The 2½" *playpipe*, a nozzle with two handles and smoothbore stacked tips, has been and continues to be a popular high-flow handline nozzle. As previously mentioned, the tip sizes and their respective flows according to traditional operations are the following: 1" at 50 psi nozzle pressure flowing 210 gpm; 1⅛" at 50 psi nozzle pressure flowing 265 gpm; and 1¼" at 50 psi nozzle pressure flowing 325 gpm. Now let's see what we can get from these tips at higher nozzle pressures (table 9–7).

To simplify the calculation for the pump discharge pressure (PDP) for a 2½" handline nozzle with the commonly utilized triple-stack smoothbore tips (1", 1⅛", and 1¼"), here is an example of what can be done. Determine the maximum flow for the 2½" handline evolution based on the largest tip, which is the 1¼" tip. The PDP should reflect that flow (table 9–8). This same PDP should be used for the 1⅛" and 1" tips as well. It's that simple. One PDP is good for all three tips. Below is an example of the flows that can be delivered from the three tips in the triple-stack

Table 9–6. Kinking test

Nozzle Type/Size	Nozzle Pressure (NP)/ Nozzle Base Pressure (NBP) (psi)	Flow Rate (gpm)	Flow Rate, Kinking (gpm
TFT Automatic	100 NBP	150	105
1"	45 NP	200	170
1⅛"	40 NP	240	220
Vindicator Heavy Attack	50 NBP	250	225

Table 9–7. Flow rates from nozzle tips at varying pressures

Nozzle Tip Size	Nozzle Pressure (psi)	Flow Rate (gpm)
1"	72	252
	100	297
	120	325
1⅛"	64	301
	75	326
	87	351
	115	403
1¼"	57	350
	80	415

Table 9–8. Triple-stack smoothbore tips and flow rates

Tip Size	Flow Rate (gpm)
1¼"	325
1⅛"	295
1"	260

tips. The PDP used for these flows is based on the 1¼" tip flowing 325 gpm (fig. 9–3). (See chapter 2 for a more detailed discussion of net pump discharge pressure.)

To continue our discussion of the five rules, with regard to rule #3, I am not sure why the tip diameter can be no larger than one-half the size of the diameter of the hose. IFSTA offers no explanation for this. Many departments have been successful with nozzle/hose combinations that break this rule. In fact, the FDNY is well-known for their highly successful interior attack handline, which uses the 15⁄16" tip and 1¾" hose. I have had success with 1", 1⅛", and 1¼" tips using 1¾" hose as well.

Now let's go back to rules 4 and 5 for smoothbore tips used with handlines. Rule 4 from IFSTA states that when operating a smoothbore handline, the ball valve must be fully open to produce the best quality stream. A partially opened valve will create turbulence in the waterway, breaking up the stream. It should be fully opened or fully closed. This rule is correct in regard to getting maximum flow, and it is generally a useful practice to follow. If the nozzle needs to be gated down for safety, however, I have found that most smoothbore tip nozzle combinations can be gated down to nearly the halfway point, reducing the nozzle reaction enough to maintain safe control of the nozzle while still providing a working stream. The stream will likely be less smooth, but the key is getting "the wet stuff on the red stuff." Does the stream have the reach? Is it hitting the target? This is

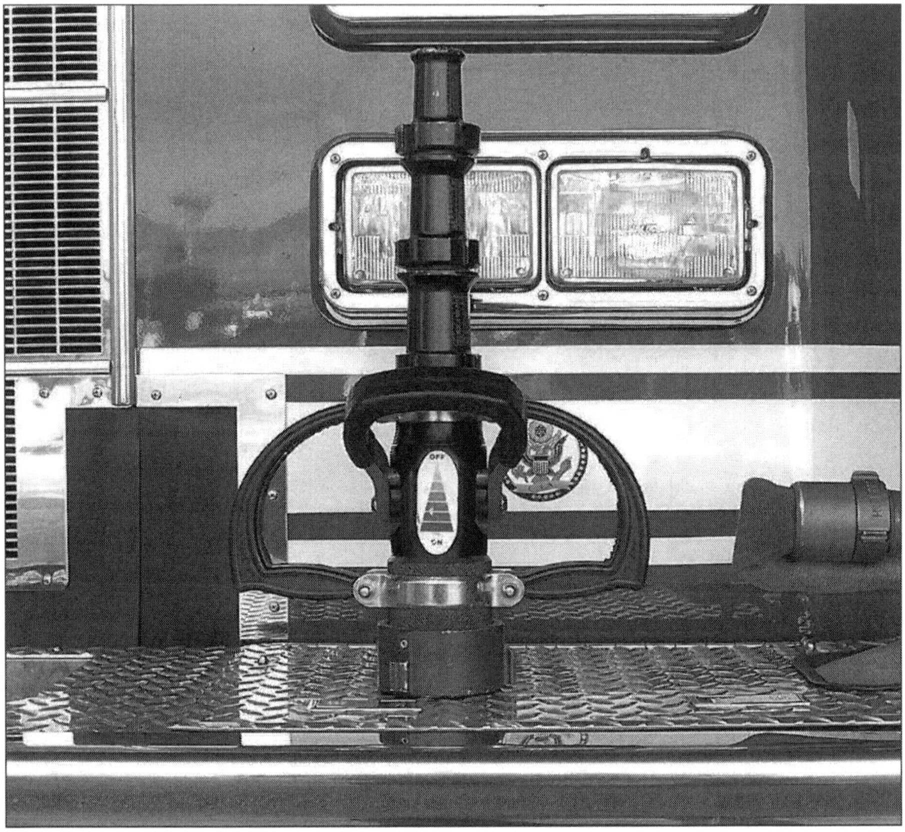

Figure 9–3. The triple stacked tips (1", 1⅛", and 1¼") on a 2½" or 3" handline can provide a variety of flows using nozzle pressures higher than 50 psi. The only thing that is required is to think outside of the box.

what we should be looking for. You need to evaluate the stream based on performance, not appearance. Firefighters tend to do the opposite.

The following flow tests (table 9–9) are designed to show 1¾" handline nozzle capabilities. The first series of tests will show how far an interior attack nozzle can be gated down and still provide a working stream. The streams were analyzed at a point when they either reached a minimum of 25', had a flow of 100 gpm, or broke up and became inefficient, whichever came first.

The next series of flow tests (table 9–10) shows high-flow handlines first in a maximum flow operation with the bale of the nozzle fully opened and then gated down to a point where the stream had a minimum reach of 50', reached a flow of 250 gpm, or broke up and became inefficient, whichever came first.

It should be noted that a 1½" stream shaper can be used with the handline/smoothbore tip combination to clean up the stream much as a master stream

Table 9–9. Smoothbore flow tests, 1¾" handline, fully opened vs. gated down

Smoothbore Nozzle Size	Maximum Flow (gpm)	
	Fully Opened	In Gated Position
15/16"	185	145
1"	210	170

Table 9–10. High-flow handlines maximum flow, fully opened

Nozzle Tip Size	Nozzle Pressure (psi)	Flow Rate (gpm)
1"	72	252
	100	297
	120	325
1⅛"	64	301
	75	326
	87	351
	115	403
1¼"	57	350
	80	415

shaper is used. It should be placed between the nozzle valve and the tip. The small shaper has also provided improvement to nozzles using the slug tip, which is built into the nozzle valve. For this application, placing the shaper between the hose and the nozzle works best.

The final rule we will discuss from the IFSTA guidelines is rule 5, which states that the size of the waterway in the ball valve should be larger than the tip size. If the waterway is smaller than the tip size, it will act as the tip itself, thus creating turbulence as it enters into the larger tip size. This will produce a broken stream. As stated previously, this rule is accurate.

High-Pressure Master Stream Operations Using Smoothbore Tips

A *master stream* is a heavy-caliber stream delivered through a master stream water delivery appliance. A master stream is used when flows surpass 350 gpm and become too difficult to deliver from a handline operation due to nozzle reaction. The stream that a master stream operation produces is high in flow and is usually a straight stream using a smoothbore tip or some type of combination nozzle.

There are three types of master stream operations: the *fixed master stream*, the *portable master stream*, and the *elevated master stream*. The following information about master stream smoothbore tips can be found in most fire stream management books in circulation today. It is foundational information for establishing effective operations.

The most common set of smoothbore tips that comes with a master stream is a set called the *stacked tips*. The set consists of four tips connected into one stack, which is then attached to the master stream appliance. The tip sizes and the corresponding flows are based on 80 psi nozzle pressure (table 9–11).

For master stream appliances capable of flows higher than 1,000 gpm, the following tip sizes with 80 psi nozzle pressures will apply. Again, these are the most common (table 9–12).

It should be noted that the conventional wisdom relating to smoothbore tip nozzle pressures used with master stream operations is to use a nozzle pressure of 80 psi. Nozzle pressures higher than this will supposedly create a broken and insufficient stream, as well as an unstable operation in the elevated and portable modes according to most fire stream books. Conventional wisdom also dictates that nozzle pressures up to 100 psi can only be used with fixed master stream operations. Nevertheless, we will demonstrate how smoothbore tips can be used to produce high-pressure streams beyond these conventional parameters while remaining safe and efficient in all modes of operation.

Table 9–11. Stacked smoothbore tips with nozzle sizes and corresponding flow rates at 80 psi nozzle pressure

Nozzle Size	Flow Rate (gpm)
1⅜"	502
1½"	598
1⅜"	814
2"	1,063

Table 9–12. Common stacked tips for flows higher than 1,000 gpm at 80 psi nozzle pressure

Nozzle Size	Flow Rate (gpm)
2¼"	1,345
2½"	1,661
2¾"	2,010

The purpose of high-pressure smoothbore tip operations used in conjunction with a master stream is to provide the required water flow rates and a high velocity stream that improves the reach and penetration capabilities of the firefighting stream while still maintaining efficient stream performance.

The maximum allowed nozzle pressure for this type of operation is based on the following rules:

1. *Maximum allowed inlet pressure to the master stream appliance.* Manufacturers of master stream appliances for the municipal fire service include Task Force Tips, Akron Brass, and Elkhart Brass. The specifications are the same for each of the three brands of these appliances in the fixed mode:
   - 200 psi inlet pressure
   - 631 lbs NP at 1,250 gpm
   - 1,010 lbs NP at 2,000 gpm

   You will find that most of the other master stream manufacturers require 200 psi inlet pressure as well.

   Note: Never exceed the inlet pressure when performing high-pressure operations.
2. *Maximum allowed flow for the master stream appliance.*
3. *Maximum allowed nozzle reaction for the master stream appliance.* This is based on a 100 psi combination nozzle at the rated flow of the appliance. For example, a master stream appliance using a 100 psi combination nozzle flowing 1,250 gpm has a nozzle reaction of 631 lbs of reactive force.
4. *Maximum allowed operating pressure for the discharge hose supplying the master stream appliance when applicable.* This applies to portable master stream operations and elevated stream operations using a non-quint truck company.
5. *Stream angle capabilities for the appliance/apparatus combination.* In order to provide the required pressure for high-pressure operations, the pump operator needs to understand the limitations mentioned above and throttle up accordingly. Flow tests will determine what the maximum pressure can be, again based on the previously mentioned criteria. A pump chart should then be designed showing all possible pump operations that can be used. The pump operator should plan on throttling up to the maximum allowed predetermined pressure unless other factors prevail to prevent doing so. Does this mean that every time a deck gun operation is placed into service, the pump discharge pressure needs to be maxed out? No, but the predetermined pump

discharge pressure is the number that the pump operator will try to reach unless redirected by instruction or by circumstances.

Factors that may stop the operation from going to maximum pressure include the following:

a. Insufficient water supply
b. Insufficient throttle capability
c. Insufficient engine speed capability
d. Goal accomplished/fire extinguished before maximum pressure is reached
e. Rules set by the department
f. Orders from the company officer

High-Flow, High-Velocity Operations

The following smoothbore tip sizes with corresponding nozzle pressures and flows have successfully produced high-velocity/high-flow streams using a master stream (table 9–13). The maximum nozzle reaction is 631 lbs for a 1,250 gpm appliance and 1,010 lbs for a 2,000 gpm appliance.

Low-Flow, High-Velocity Operations

Picture a large fire that has totally consumed a structure, leaving a large pile of burning rubble to be overhauled. This is where the digging power of a high-velocity stream can be very effective. The smoothbore tip sizes and corresponding nozzle pressures and flows given in table 9–14 have been successfully used in this type of operation. One point to remember is that the goal in this type of operation is to produce velocity, not high flows (fig. 9–4).

Working a brush fire from a fixed master stream appliance is much easier than pulling hose. When possible, a fixed master stream can make far-reaching sweeps in burn areas with the above-mentioned high-pressure stream operations. In fact, the $1\frac{1}{8}$" tip application will likely be the tip of choice for this operation.

These same three flow operations also apply to portable master streams and elevated master streams. With elevated master streams, however, you must follow the guidelines for flowing water at specific angles (table 9–15). This information is always found on a placard on the turntable console.

Table 9–13. High-flow/high-velocity master streams with nozzle reactions

Tip Size	Nozzle Pressure (psi)	Flow Rate (gpm)	Nozzle Reaction (lbs reactive force)
1⅜"	175	743	519
1½"	150	819	530
1¾"	120	997	577
	150	1,114	721
2"	100	1,188	628
	110	1,246	691
	120	1,302	754
	140	1,406	879
	150	1,455	942
	160	1,503	1,005

Note: The 2" tip at 160 psi nozzle pressure produced a stream with a reach of 360'.

Table 9–14. Smoothbore tip sizes and corresponding nozzle pressures for low-flow, high-velocity operations

Tip Size	Nozzle Pressure (psi)	Flow Rate (gpm)
1⅛" tip up to	180	504
1¼" tip up to	180	623
1⅜" tip up to	180	754

Figure 9–4. This deck gun uses a 1¼" smoothbore tip flowing 623 gpm at 180 psi nozzle pressure.

Table 9–15. Data for 2,000 gpm appliance and waterway on an elevated platform with a 2" tip

Nozzle Pressure (psi)	Flow Rate (gpm)	Reach	Nozzle Reaction (lbs reactive force)
160	1,500	220'	1,005

Developing the Required Pump Discharge Pressure for the Portable Monitor

The following information is based on any size discharge hose used as either a single line or siamese lines. Just as with the fixed master stream operation, the portable monitor can deliver several different flows, with each potentially having a different pump discharge pressure. It is not practical to list all of them on the pump chart. The only numbers needed are for the maximum flow operation, with the most common being 1,250 gpm with a combination nozzle at 100 psi nozzle pressure. Adhere to the specific manufacturer's ratings on what the highest flow in the portable mode should be. The manufacturer will usually also indicate what the friction loss in the appliance will be at the high flow rate. Add the appliance friction loss and the nozzle pressure together and note it on the chart. The number for the nozzle pressure should be based on the highest nozzle pressure used in the department's arsenal.

If a department uses smoothbore tips rated at 80 psi and a 100 psi combination nozzle for portable monitor operations, the pump chart should reflect the higher 100 psi figure in the formula listing on the pump chart. The friction loss per 50' or 100' of the hose evolution being used (whichever works best) should also be noted. Remember, the friction loss is based on the highest flow allowed by departmental operations. Determine how much hose is being used in the evolution and add up the friction loss plus the numbers from the appliance and nozzle pressure to develop the pump discharge pressure. Consider an example with a portable monitor that is flowing 1,000 gpm through an automatic nozzle with a 300' length of 5" discharge line. Add 15 psi for the appliance friction loss, 100 psi for the nozzle pressure based on the fact that the highest nozzle pressure used is 100 psi, and 5 psi per 100' of 5" hose (15 psi for the 300'), bringing the sum to 130 psi.

When using smoothbore nozzles, the maximum flow pump discharge pressure listed on the pump chart will work with any size tip from 2" down to 1" while still providing for safe and efficient operations (fig. 9–5). It is true that the rated flow at the standard smoothbore nozzle pressure (80 psi) will probably be surpassed. However, it will do this while keeping within the required inlet pressure of the appliance (200 psi) and the maximum allowed flow and nozzle reaction. Consider another example in which the pump chart lists a pressure of 115 psi plus 5 psi per 100' of 5" for a portable master stream supplied by 5" LDH flowing 1,000 gpm. The 115 psi represents the appliance friction loss of 15 psi and the nozzle pressure for the automatic nozzle, which is 100 psi. Now, for whatever reason, the automatic nozzle is being replaced with the 1⅜" tip. The pump discharge pressure for the 1⅜" tip at its rating of 500 gpm has a friction loss of 2 psi for the 5" hose. But since the pump discharge pressure will remain at the high-flow pressure, the excess pressure in the hose (3 psi per 100') will be added to the nozzle pressure, creating a slightly elevated flow. A picture is worth a thousand words, so do the flow tests and note the nozzle pressure readings.

Remember to pump to the largest tip. Given the data in table 9–16, the flow rate and nozzle pressure will be based on a 2" tip flowing 1,060 gpm at 80 psi nozzle pressure (fig. 9–6).

Figure 9–5. Flow rate of 1,500 gpm at 160 psi nozzle pressure

Table 9–16. Smoothbore tip sizes with corresponding flow rates and nozzle pressures for a portable monitor stream.

Tip Size	Flow Rate (gpm)	Nozzle Pressure (psi)
1⅜"	589	110
1½"	634	90
1¾"	814	80
2"	1,060	80

Figure 9–6. This is a set of stacked tips for a 1250 GPM master stream. The tips are $1\frac{3}{8}$", $1\frac{1}{2}$", $1\frac{3}{4}$", and 2".

Fixed-Gallonage Combination Nozzle

A combination nozzle is designed to produce a straight stream and a fog pattern varying from narrow to wide. A fixed-gallonage combination nozzle works just like the smoothbore nozzle in regard to its flow/pressure rating. It is rated to produce a specific flow rate at a specific nozzle base pressure. The internal part that creates the flow rating is a baffle (disc) located at the circular opening of the nozzle. The baffle works by being set either closer to the opening, which lowers the flow rate, or further away, increasing the flow rate.

There are two types of combination fixed-gallonage nozzles. One is actually called the *fixed gallonage nozzle* because the baffle is permanently set in one position. The other is the *selectable gallonage nozzle*, which allows the baffle position to be selected by turning an adjustment ring that moves the baffle position. Both of these nozzles can be overpumped or underpumped to change the rating by changing the nozzle base pressure to some degree.

The Vindicator Nozzle

When it comes to extinguishing fires, water is used almost exclusively; there are few exceptions. The method of delivery has always been through discharge evolutions made up of hoselines, appliances, and nozzles. I like to think of the nozzle as an applicator for applying the water for fire extinguishment.

Over the years there have been several types of nozzles used for firefighting, with various results. The Vindicator nozzle developed by First Strike Technologies, although not brand new, is one of the more recent nozzles to enter into the fire service (fig. 9–7). It is not a combination nozzle or a smoothbore nozzle. Some people describe it as a large bore nozzle. It varies from a smoothbore nozzle in many ways.

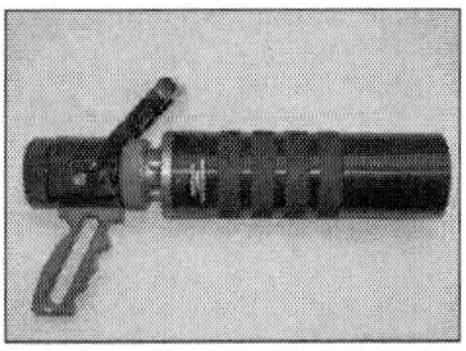

Figure 9–7. The Vindicator nozzle consists of two parts: the ball valve and the barrel. The ball valve has a 1⅜" waterway with flow capabilities up to 500 gpm. The master stream nozzle flows up to 1,000 gpm.

The Vindicator nozzle is a fixed-gallonage nozzle in that it delivers a specific flow at a specific nozzle base pressure. It has a flow range that delivers a workable stream varying from low to high with an accompanying nozzle base pressure. The nozzle base pressure varies from 40 psi to 100 psi, offering a wide flow capacity range and optimal stream performance for the Heavy and Blitz Attack models.

The Vindicator nozzle only provides a straight stream. The nozzle consist of two parts, the ball valve and the barrel. The ball valve has a 1⅜" waterway with flow capabilities up to 500 gpm, while the master stream nozzle flows up to 1,000 gpm. Inside the barrel there is a stream shaper and a stream deflector that develop the water into a workable firefighting straight stream that offers lower nozzle reaction characteristics. The handline barrel portion of the nozzle utilizes 1½" couplings, while the master stream has 2½" couplings. The barrel has air intake ports at the base, similar to a foam aspirated nozzle, which are designed to introduce air into the fire stream, creating large droplets of water.

There are four types of Vindicator nozzles:

1. *Light Attack.* The Light Attack nozzle has a flow range of 90 gpm to 200 gpm, and a nozzle pressure range of 50 psi at 95 gpm *and* 100 psi at 200 gpm.
2. *Heavy Attack.* The Heavy Attack nozzle has a flow range of 175 gpm to 425 gpm, and a nozzle pressure range of 50 psi at 250 gpm *and* 100 psi at 425 gpm.
3. *Blitz Attack.* The Blitz Attack nozzle has a flow range of 200 gpm to 500 gpm, and a nozzle pressure range of 50 psi at 325 gpm *and* 100 psi at 500 gpm.
4. *Master Attack.* The Master Attack nozzle has a flow range of 675 gpm to 1,000 gpm, and a nozzle pressure range of 65 psi to 85 psi (fig. 9–8).

The following flow tests were conducted by the NYFD to compare the performance capabilities of a Vindicator Heavy Attack nozzle and a 15⁄16" smoothbore tip nozzle. The tests included flow delivery, nozzle reaction, and flow impact.

Figure 9–8. The Vindicator Master Attack nozzle flowing 1,000 gpm

Vindicator Heavy Attack Nozzle Flow Rate and Nozzle Reaction Tests

Flow test

The flow test for water delivery was based on the nozzle reaction of the 15⁄16" smoothbore tip at 180 gpm. The nozzle reaction was 62 lbs of reactive force. The purpose of this test was to approach as nearly as possible the 62 lbs of nozzle reaction force with the Vindicator nozzle and then register the flow rate at that point (table 9–17).

The Vindicator nozzle was able to flow 60 more gpm at the same basic nozzle reaction as the 15⁄16" smoothbore tip flowing 180 gpm.

Impact test

This test was done using an impact plate attached to a certified load cell that measures actual impact in pounds per square inch. This test is meant to show which nozzle has the best penetration capabilities based on the power of the stream. Both nozzles were tested at their respective flows (table 9–18).

Let's take a look at some nozzle comparisons to see how the Vindicator stands up (table 9–19).

A comparison of the nozzles reveals that the Vindicator nozzle generally offers the flexibility of an interior attack line with flows reaching the capabilities of a 2½" line.

Table 9–17. Vindicator nozzle vs. 15⁄16" smoothbore nozzle flow test

Tip	Flow Rate (gpm)	Nozzle Reaction (lbs reactive force)
15⁄16" Smoothbore	180 gpm	62
Vindicator Heavy Attack	240 gpm	60

Table 9–18. Vindicator Heavy Attack nozzle vs. 15⁄16" smoothbore nozzle impact test

Tip	Flow Rate (gpm)	Impact Measurement (psi)
15⁄16" Smoothbore	180 gpm	53.4
Vindicator Heavy Attack	240 gpm	60.2

Table 9–19. Nozzle pressures and nozzle reactions for Vindicator Heavy Attack nozzle vs. smoothbore and combination nozzles

Flow Rate (gpm)	Nozzle	Nozzle Pressure (NP)/Nozzle Base Pressure (NBP) (psi)	Nozzle Reaction (lbs reactive force)
150	Combination	100 NBP	76
250	Combination	100 NBP	126
250	Combination	50 NBP	89
210	1" Smoothbore	50 NP	79
265	1⅛" Smoothbore	50 NP	99
240	Vindicator Heavy Attack	45 NBP	60

Foam capabilities

As mentioned previously, the air inlet ports at the base of the barrel help create the large droplets of water in the fire stream. Besides this capability, the air inlet ports also let you use the Vindicator as a foam nozzle. Depending on the foam manufacturer, you can expect foam expansion ratios ranging from 7 to 1 up to 15 to 1 without a reduction in stream performance.

The Automatic Nozzle

The most debated piece of equipment and the one that elicits more passion from firefighters than anything else has been the nozzle. Let's focus on one of the most controversial—the automatic nozzle. I will discuss what the automatic nozzle is designed to do and how it can be used.

History

The automatic nozzle was invented by Clyde McMillan, the founder of Task Force Tips and former fire chief of the Gary, Indiana, Fire Task Force. The Gary Fire Task Force was a volunteer fire company formed to assist the municipal fire department in Gary on large structure fires. For the most part it was a big water company that mainly provided master stream operations.

Large volume water delivery fires can have water supply issues that negatively affect the quality of the fire streams. There were two basic types of master stream nozzles used in those days, a smoothbore tip and a combination nozzle. Both of these nozzles are considered to be fixed gallonage because they require a specific nozzle pressure to produce the rated flow of the nozzle. This in itself can be difficult to achieve at large flow fires with hydrant system pressures dropping as the fire flow demands increase. Chief McMillan was well aware of this problem and decided to come up with a solution that would provide a good quality stream as far as reach is concerned, even with an uncertain water supply. Thus the automatic nozzle was born.

How it works

An automatic nozzle is designed with a pressure control unit that is calibrated to the required nozzle pressure, thus maintaining a near-constant pressure throughout the flow range of the nozzle. The reason why it is termed *near-constant pressure* is because there is a slight difference in the nozzle pressure from the bottom of the flow range to the top of the flow range of the nozzle. *NFPA 1962: Standard for the Care, Use, Inspection, Service, Testing, and Replacement of Fire Hose, Couplings, Nozzles, and Fire Hose Appliances* states that an automatic nozzle can have a nozzle pressure range of plus or minus 15% within the flow range of the nozzle.

Here is a simpler way to explain this function using a garden hose. Most people have used a garden hose without a nozzle connected to it. It is simply an open butt hose, which is effective for filling things up. If you need to get any reach with a garden hose stream, however, you put your thumb over the opening of the hose. This creates back pressure and allows you to project a decent water stream for the task at hand. Consider how you would use your thumb to maintain a water stream to reach a flowerpot 15" away. You would likely depress your thumb into the stream to create more reach, and you might retract your thumb if the stream went past your target. This is exactly the way an automatic nozzle works, except it does not require input from the operator. It works automatically based on its design.

Theory

It is important to understand that the automatic nozzle concept is based entirely on maintaining a near-constant nozzle pressure and thus producing the best possible stream. Everything else about delivering water through a handline or master stream with an automatic nozzle is no different than delivering water to a fixed-gallonage nozzle. You still need to have enough water in the water supply to support the required flow rate from the nozzle. You still have to calculate friction loss for the required flow in the hose and/or appliance. The big difference with the automatic nozzle is that if the flow rate requirement for the fire cannot be achieved, as long as the flow rate is within the flow range of the automatic nozzle, the stream reach quality will for the most part be maintained. With a fixed-gallonage nozzle, the stream reach will drop when the required pressure for the nozzle rating cannot be met, whether the drop is from a wrong-sized tip, a depleted water supply, or just an incorrect engine pressure (fig. 9–10).

Table 9–20 shows two types of fixed-gallonage nozzles, a combination nozzle and a smoothbore nozzle. It shows how a drop in nozzle pressure reduces the flow. What the chart does not show is that when the nozzle pressure decreases, so does the reach of the stream.

Automatic nozzles solve this problem

When smoothbore tips are being used and the rated flow of the nozzle cannot be produced, firefighters can reduce the tip size to improve stream quality. It will take time and a little bit of guesswork to make this work, however. With the automatic nozzle there is no delay or guesswork. The nozzle adjusts to the flow being delivered to it.

One problem with a fixed-gallonage combination nozzle is that there are no other choices to switch to. You have to use the stream it is producing, even if you cannot increase the water flow rate to the rated flow of the nozzle.

Applications

Automatic nozzles have two applications—handlines and master streams. The flow rates for the small handline nozzles range from 90 gpm to 200 gpm depending on the model. Nozzle pressures for the small handlines can vary from 55 psi to 100 psi, once again based on the model. They are usually found on 1¾" interior attack lines but can be used on booster lines, 1½" lines, and 2" lines. The big 2½" handlines have a flow range from 250 gpm all the way up to 500 gpm and also have nozzle pressures ranging from 55 to 100 psi. For municipal applications, master stream nozzles can vary from 500 gpm to 2,000 gpm, with nozzle pressures ranging from 75 psi to 120 psi based on the model.

Unlike smoothbore nozzles, automatic nozzles have internal components that could allow debris to interfere with the stream quality. A flushing mechanism is built into automatic nozzles and does an excellent job of flushing the debris. Flushing can be done by turning the nozzle all the way to full fog and then going just a little further until the rotation stops, which will put the nozzle in the flush mode. It is important to gate the nozzle down before getting out of the flush

Figure 9–10. The automatic nozzle fixes the fire stream performance problem by adjusting to the available flow rate within its flow range, while keeping a near-constant nozzle pressure. If personnel cannot obtain the rated flow rate for a fixed-gallonage nozzle for any reason, the nozzle pressure will be below its rating, and the reach of the stream will be diminished.

Table 9–20. Nozzle pressures and corresponding flow rates for fixed-gallonage nozzles

Fixed-Gallonage Nozzle	Nozzle Pressure (psi)	Flow Rate (gpm)
Combination	100	1,000
	90	949
	80	837
2" Smoothbore	80	1,063
	70	994
	60	920
	50	841

mode because of the possibility of nozzle reaction kicking in and creating an unsafe situation.

Manufacturers of automatic nozzles have guidelines for field maintenance of the nozzles, including the use of specific lubricants.

True or False: Nozzle Questions

We know that firefighters are passionate about their nozzles, and there has been much debate comparing automatic nozzles to fixed-gallonage nozzles. It seems to me that much of the information about automatic nozzles has been false, or at least misunderstood. Let's take a look at some of the topics of debate about automatic nozzles and analyze them.

1. The automatic nozzle can be unsafe because it will always give you the appearance of a good stream, but it may be misleading if it is lacking the required flow rate.

   If you are relying only on stream appearance to determine whether your handline is flowing the required rate or not, then I guess this would be a true statement. A firefighter should have a good idea of what the nozzle reaction feels like for the line he or she is using and should be able to determine to a degree if the stream is underpumped. Let's take this one step further. In an interior attack, how far does a stream have to reach? The fact is that any nozzles, not just automatic nozzles, can have underpumped streams, and yet the streams will appear to be sufficient in the confined spaces of an interior attack in a structure. In interior operations, there is a good chance that you will not be able to see the entire stream because of the fire and associated smoke.

2. A smoothbore nozzle has a harder hitting stream than an automatic nozzle.

   This is another myth. The truth is that the highest nozzle pressure will have the highest or hardest impact. Many flow tests have proven this.

   Here's a simple flow test you can conduct to prove this point. Place an automatic nozzle and a smoothbore nozzle side by side, pointing straight up in the air, and connect them to portable master streams. The smoothbore nozzle should have $1\frac{1}{8}$" tip and be pumped at 50 psi nozzle pressure to achieve a flow rate of 265 gpm. The automatic nozzle will be pumped at 100 psi nozzle pressure and will also flow rate of 265 gpm. To make this test accurate, a flowmeter should be used for the automatic nozzle and a handheld pitot gauge should be used to measure the nozzle pressure for the smoothbore nozzle. This test is shown in the sequence of photos in figure 9–11. As you can see, especially in the center photo, the streams look identical. The photo on the right shows the maximum height of each stream, with the automatic nozzle stream going approximately 10' higher than the stream with the smoothbore nozzle. We used gravity for resistance.

3. The automatic nozzle is a cure-all for a poor water supply.

   This is absolutely false. Automatic nozzles cannot create water; water has to come from the water supply. The automatic nozzle can create a decent stream even with a low water supply, but it is important to understand that a decent stream in terms of reach does not necessarily have the required water flow rate.

Figure 9–11. Automatic and smoothbore tip flow tests

10

THE 2½" HANDLINE

Introduction

All of the nozzles mentioned in this book can be used on a 2½" handline with varied results. The 2½" handline has been used in the fire service for decades. My research suggests it was the first and only handline used when fire hose was introduced into the fire service. Today the 2½" handline is known for its large flow capabilities for firefighting in both offensive and defensive modes. Most engine companies have a preconnected 2½" line designed specifically for a quick attack. Static hose beds are often set up with the end of a 2½" hose already attached to a nozzle for a quick attack. These handlines are also used to surround and drown fires. In recent years, use of the 2½" handline has increased for two reasons. First, firefighters are realizing the importance of large flows to achieve quick knockdowns. Second, new hose handling techniques have been developed that utilize body mechanics for hose handling instead of brute strength.

Some nozzles offer lower nozzle pressures, while others go with the standard 100 psi nozzle pressure. Both have been used for years. Improvements continue in the use of smoothbore nozzle tips, and now there is a 1$^{3}/_{16}$" nozzle tip that offers a flow of 300 gpm at 50 psi nozzle pressure. The Vindicator nozzle, as well as larger smoothbore tips and larger combination nozzles, also have increased flows up to 500 gpm. Obviously these larger flows are not made for advancement into a structure. Instead they are used for a stationary sit-down hit from the outside to knock the fire down. The advancing 2½" handlines are usually armed with 1$^{1}/_{8}$" tips, with flows varying from 250 to 265 gpm.

The quality of the 2½" hose has changed over the years, primarily making the hose lighter. The thing to remember about lighter hose is that in order to make it lighter, less material is used in its construction, which in turn makes it more susceptible to abrasions and kinking.

There are other uses for the 2½" hose that do not involve handlines. It may be used as a discharge line into either an appliance or a fire department connection in a multistory building. High-pressure high-rise hose is now being manufactured that has working pressures of 600 psi.

The Big Paulie Nozzle

The 2½" handline has always been known for its high flow rates. In fact, there are many different flow rates that are being used successfully by fire departments nationwide with this big line. The focus in this section will be on a new nozzle concept called the *Big Paulie nozzle*. All flows listed, however, can be used on standard smoothbore nozzles, the Vindicator nozzles, and some combination nozzles.[1]

The Big Paulie nozzle is not a large-caliber stream being delivered from a master stream appliance. It is a nontraditional handline nozzle used for a heavy attack from a stationary position (fig. 10–1). A 500 gpm stream is a substantial amount of water, more than capable of getting the big hit on a large fire as you would expect from a standard 2½" handline flow. The reason 500 gpm was chosen is twofold. The 500 gpm handline evolution is at maximum flow in regard to what two firefighters can handle based on sitting on the line in a stationary mode. Flow tests have proven that two firefighters can safely and efficiently deploy a 500 gpm handline using the techniques that will be presented in this section. Flows higher than this are more difficult to handle and are not recommended as handline operations.

The Big Paulie handline consists of four parts:

1. *A 2½" nozzle valve.* It has a 2" waterway specifically designed for large flows.
2. *A 2½" stream shaper.* The stream shaper is 10' in length and is designed to reduce turbulence as water enters the smoothbore tip, as well as to provide leverage for the firefighter directing the stream.
3. *The stream-directing handle.* The handle is placed between the stream shaper and the smoothbore tip. The handle swivels, which makes it easier for the firefighter to direct the stream. This handle also makes it easier to move the handline when it is not flowing.
4. *The nozzle.* The nozzle is either a 1⅜" or 1½" smoothbore tip, 9" in length, which is capable of a flow rate of 500 gpm at 80 or 55 psi nozzle pressure, respectively.

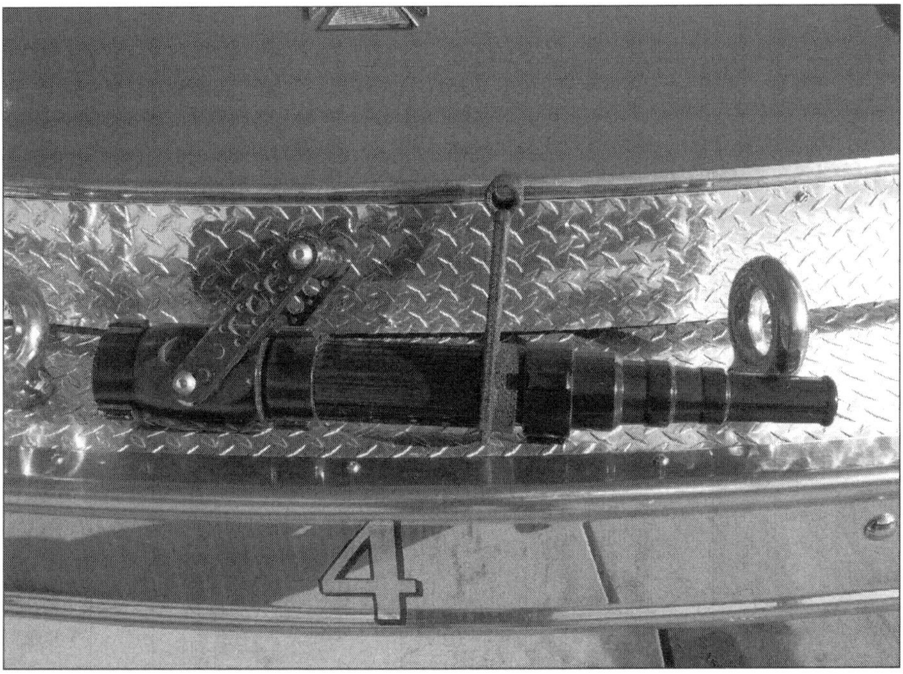

Figure 10–1. The Big Paulie handline consists of four parts assembled together to make up the nozzle.

Nozzle-Handling Techniques

The nozzle-handling technique for the Big Paulie high-flow handline is simple and is a modified version of a technique that has been used for decades. Just have a seat. That's right, it is the same technique used for sitting on a 2½" line with a loop, with two modifications. First, the loop is not needed. Numerous flow tests have proven that the loop does not provide much additional stability. If you prefer to make a loop, however, that is entirely acceptable. Second, a minimum of two firefighters are required to sit on the line because of the nozzle reaction. The firefighter at the sits directly on the hose, about 2½" to 3' from the nozzle. The entire handline stays on the ground except for the last 3' that the firefighter is handling. The hose on the ground is important with this concept because the ground actually absorbs most of the nozzle reaction. As mentioned, looping the hose at this point makes little difference as to how well the nozzle reaction is absorbed. Furthermore, in a quick attack/minimum manpower situation, setting the loop up takes time away from the initial quick hit, especially if the line is already charged.[2]

The second firefighter sits directly on the hose behind the firefighter on the nozzle (fig. 10–2). It is extremely important that both firefighters keep all their weight on the hose while flowing water to keep the nozzle reaction at a minimum. It is also equally important for the firefighter at the nozzle to keep one hand on the bale of the nozzle at all times in case it needs to be gated down to be manageable or shut down completely.

If you decide that the 500 gpm flow from a 1⅜" nozzle tip is too much (with a nozzle pressure of 80 psi and a nozzle reaction of 237 lbs), a 1½" tip at 55 psi nozzle pressure with a nozzle reaction of 194 lbs will also flow 500 gpm with less effort. The trade-off will be less reach, but definitely still enough reach to hit most targets (fig. 10–3).

Figure 10–2. The backup firefighter sits on the hose to help support the nozzle operator on a Big Paulie high-flow handline.

Figure 10–3. The high-flow handline stream resembles the quality of a master stream.

There is also an entire range of 2½" flows less than 500 gpm that can be highly effective. The following list contains selected tips with their corresponding nozzle pressures and flow rates from 210 gpm up to 500 gpm (table 10–1).

Table 10–1. Tip sizes and corresponding nozzle pressures for flow rates of 500 gpm or less.

Tip Size	Flow Rate (gpm)	Nozzle Pressure (psi)
1"	210	50
1⅛"	265	50
1¼"	325	50
1"	300	100
	325	120
1⅛"	300	64
	326	75
	351	87
1¼"	350	57
	413	80
1⅜"	400	50
	500	80
1½"	500	55

A Small Line with a Big Punch

High-performance 1¾" and 2" hose

Modern fire departments typically have the ability to deploy 2½" handlines. They often equip the big line as a preconnect to make a quick large-flow attack on a significant fire to achieve a rapid knockdown. This is usually done by the first-in unit. Of course, they also have the ability to make up 2½" handlines from static loads in the hose bed for large flow operations. The flows for these big lines usually range from 250 to 325 gpm. When it comes to large flow handlines, the 2½" hose has always been the weapon of choice since its lower friction loss capabilities allow for more flow.[3]

The drawback to using 2½" line is that it is heavy, which makes it more difficult to deploy, especially in a first-in quick attack mode. It is also fairly difficult to move around after it is charged. Because of these drawbacks, firefighters tend to avoid using the 2½" line. As a result, smaller lines are often pulled instead, and lower-than-required flows are delivered.

Repeating history

It's a true fact that the fire service tends to do what they have always done in the past when it comes to certain issues. The 2½" handline is a good example of this. History tells us that it is the only line that works for big flows and for high-rise fires. Assistant Chief Bill Nemick, Pueblo, CO, Fire Department, made an interesting point in an article he wrote for *Fire Engineering*. He stated that firefighters tend to choose the size of the attack line based on the type of fire.[4] A 1¾" line is used for small fires such as single family dwellings, and the 2½" line is used for the large commercial fires. Once again, this strategy is based on what has always been done in the past. Assistant Chief Nemick recommended that we match the water flow rate to the heat energy (Btus) and use the water delivery equipment that is best suited for the job.[5] As you will see, the 1¾" high-performance line and the 2" line can be used for the high-flow fires in certain situations.[6]

Another good example of doing things the way we have always done them is found with the handline used for high-rise firefighting. Again, conventional wisdom tells us that the only line that can be used is the 2½" line because of the pressure issues in high-rise structures. We are now proving that 2" and 1¾" lines are also options in some situations, as can be seen from the flow test data that follows.

High-rise fire protection systems: Handline flow tests

Flow tests were conducted at the 25-story Riviera Hotel in Las Vegas, NV on August 24, 2015, with 1.88", 2" with 2½" couplings, and 2½" hoselines, all 150' in length. All hose was provided by Key Hose. We used smoothbore nozzles with ⅞", 15/16", 1", 1⅛", and 1¼" tips to deliver the water. The flow tests were measured on the roof to show a worst-case scenario of what the pressures and flows would be. The system was rated at 500 gpm with a 65 psi standpipe outlet residual pressure (SORP). The system specifications were 500 gpm at 65 psi standpipe outlet residual pressure.[7]

The tests were conducted from the hydraulically furthest standpipe outlet from the building pump (table 10–2).

Additional flow tests were conducted with the same system specifications to compare the performance of a 2½" handline and a 1¾" high-performance handline (table 10–3). The pump discharge pressure is for a 200' handline, while friction loss is for a 100' length of hose.

Better techniques for smaller handlines

In recent years there has been a resurgence in training and advocating for the use of the 2½" attack line. Several fire service instructors specialize in training with advancing the big line with very positive results. They have demonstrated techniques that remind me of athletes training for an event. The basic use of body mechanics makes a big difference in moving the big hose.

Table 10–2. Handline flow tests with a high-rise fire protection system and varied hose sizes

(a) High-performance Combat Ready 1.88" hose × 150'

SORP (psi)	Flow Rate	Tip Size	Nozzle Pressure (psi)
85	165	7/8"	54
85	185	15/16"	50
85	210	1"	50
85	218	1 1/8"	34

(b) Eco 10 2" hose × 150' with 2½" couplings

SORP (psi)	Flow Rate	Tip Size	Nozzle Pressure (psi)
85	177	7/8"	62
85	201	15/16"	60
85	217	1"	54
80	248	1 1/8"	44

(c) Big 10 2½" hose × 150'

SORP (psi)	Flow Rate	Tip Size	Nozzle Pressure (psi)
80	265	1 1/8"	50
75	345	1 1/4"	56

Table 10–3. High-rise flow tests: 2½" handline vs. 1¾" high-performance handline

(a) Flow test: 2½" handline

Flow Rate (gpm)	Tip Size	Nozzle Pressure (psi)	Nozzle Reaction (lbs reactive force)	Friction Loss (psi)	Pump Discharge Pressure (psi)
265	1 1/8"	50	99	10	80
300	1 3/16"	50	111	15	85
325	1 1/4"	50	123	20	90

Note: Pump discharge pressure is for a 200' handline. Friction loss is for a 100' length of hose.

(b) Flow test: 1¾" high-performance handline

Flow Rate (gpm)	Tip Size	Nozzle Pressure (psi)	Nozzle Reaction (lbs reactive force)	Friction Loss (psi)	Pump Discharge Pressure (psi)
160	7/8"	50	60	20	90
185	15/16"	50	69	25	110
210	1"	50	79	35	135
265	1 1/8"	50	99	55	185
300	1 3/16"	50	111	60	190
325	1 1/4"	50	123	80	210

Note: Pump discharge pressure is for a 200' handline. Friction loss is for a 100' length of hose.

In this section we will build on the outstanding techniques taught by these instructors by making the handline lighter while maintaining the same 2½" handline flows. This section will focus on the large flow handline concept with a different twist. We will be using 1¾" high-performance attack hose and 2" hose with 2½" couplings for the high-flow handline for the purpose of making this big hit line easier to deploy.[8]

1¾" high-performance hose

The 1¾" high-performance hose has the ability to expand up to approximately 1.88" when it is charged. Aided by a friction reducing inside liner, this hose generates less friction loss and therefore allows for higher flows.[9]

Some may wonder if a 1¾" high-performance attack line will provide the flows that the 2½" line can produce (fig. 10–4). In the majority of cases, they will. Of the 2½" nozzles that are in use today, most have flow rates ranging from 250 to 325 gpm. These flow rates are achievable with 1¾" high-performance hose.

Occasionally there is a need for a handline capable of high flows for an interior attack. A 2½" line can work for an interior attack providing there is sufficient manpower and the proper techniques are used. It is not uncommon to need four to six firefighters with this scenario. It does not matter what the flow is since the weight of the 2½" line remains the same regardless of the flow rate.

Figure 10–4. Can you see the difference? The high-performance hose is only ⅛" larger in diameter.

The high-performance 1¾" hose provides a realistic alternative for a high-flow hoseline as far as ease of deployment is concerned. It is much easier to pull from the hose bed and extend when it is charged. A 50' section of high-performance 1¾" hose weighs approximately 71 lbs when charged. A 200' line weighs approximately 284 lbs charged. The same length in a 2½" hoseline weighs approximately 533 lbs.

A high-performance 1¾" high-flow line can be deployed as easily as an interior standard flow attack line with a flow of 265 gpm (fig. 10–5). Some may wonder if this handline will create manageability problems due to its rigidity and the higher pump discharge pressures than the 2½" line. To test whether or not flexibility is diminished, charge a 1¾" handline and throttle up to 250 psi pump discharge pressure, but do not flow water (fig. 10–6). Next, drag the line as you would through an obstacle course such as the burn tower, fire station, or other realistic environment. What you will likely notice is that there will be a little rigidity at the nozzle, but it should not affect the firefighter's stream-directing capabilities. The rest of the line will probably be just as easy to drag through the hallway and around the corners as if it was pumped at a common interior attack handline pressure (120–150 psi pump discharge pressure).

Since 1¾" high-performance hose has 1½" couplings, it would make sense that it could be connected to a 1¾" preconnect discharge. It is important to remember that the standard size for the small handline preconnect plumbing is 2", which will not be able to support the flows needed for a high-flow handline, at least not efficiently. Therefore, it is crucial to connect the line to a 2½" discharge, whether it is a standard 2½" discharge or one that is designed for a 2½" preconnect (figs. 10–7 and 10–8).[10]

Figure 10–5. A 265 gpm flow rate from 200' of high-performance hose

Figure 10–6. A 2½" line and the high-performance line. Here is where the high-performance hose is worth its weight in gold.

(a)

(b)

Figure 10–7. This bend is at 250 psi.

Figure 10–8. A 2½" discharge offers the best results.

One thing that was not mentioned in the flow tests was the fact that for flows above 250 gpm with the 1¾" high-performance hose, the nozzle reaction is severe enough to cause the hose to buckle right at the nozzle. Because the high-performance hose is smaller than the 2½" discharge, it does not handle the nozzle reaction as well. The 2½" hose does not have a problem with nozzle reaction at all. With that in mind, a 5' length of 2½" hose was added to the end of the 1¾" high-performance handline for the sole purpose of absorbing nozzle reaction to assist the firefighter with handling the high flows (fig. 10–9). It worked well.[11]

Safety reminder

One important safety issue to remember when using a high-performance 1¾" blitz line, or for that matter any high-flow blitz line or master stream, is to not use a smaller attack line at the same time. The reason is simple. If the big line gets shut down, the smaller line is going to take a pressure hit that could be substantial even if pressure control devices are set on the pump. The blitz line is for an initial quick hit only. It should only flow for seconds and then be shut down, to be followed by standard attack lines.

Advantages of the 2" hose

The 2" hose has been in use in the fire service long enough that it is manufactured by most hose companies. The concept behind its use is identical to the high-flow 1¾" hose previously discussed with one exception—reaching flows of 400 gpm.

There are a few advantages to the 2" hose as well as one disadvantage. The weight of 2" hose without water versus the high-flow 1¾" line (in reality about

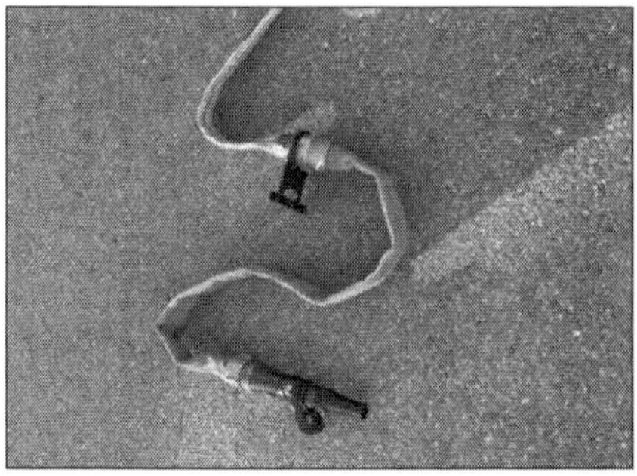

Figure 10–9. Notice the handle at the point of connection of the 2½" and 1¾" lines. This handle makes a big difference for the backup firefighter advancing the hose.

1.88") on average ranges from being the same to about 5 lbs more per 50' lengths when 1½" couplings are used. A 2" hose also comes with 2½" couplings, which will add a couple more pounds to the weight of the hose. All in all, the slight increase in dry weight is small compared to the advantages gained from its use. The difference between the two lines charged is about 12 lbs per 50'. For a 200' handline, there is a difference of 48 lbs between the high-performance 1¾" hose and the 2" hose when charged. Now compare each of these lines to the charged 2½" handline, which is nearly 40 lbs heavier per 50' than the charged 2" handline. This fact alone sells the small handline concept. Finally, most 2" hose is significantly cheaper than the high-performance 1¾" hose and is capable of the same or higher flow rates.

In table 10–4, the 2½" line is compared with the 1¾" FDNY line, the 1¾" Combat Ready line, and the 2" Combat Ready line with 2½" couplings. All lines are 200' in length. The Combat Ready hose is manufactured by Key Fire Hose. The nozzles used for this test are smoothbore tips with a nozzle pressure of 50 psi. Because the high-performance hose will have a higher friction loss than the 2½" hose at the high flows we are trying to achieve, a nozzle with a low nozzle pressure is needed to keep the overall discharge pressure as low as possible. The flows for the 2½" and 2" lines start at 265 gpm, since these are considered to be high-flow handlines. The flows for the 1¾" high-performance hose start at 160 gpm because this hose can be used for both low flow and high-flow operations. The tests include flows at 160, 185, 210, 265, 300, and 325 gpm.

As you can see, the 1¾" and 2" high-performance hoses were able to match the 2½" hose in every flow category up to 325 gpm. In fact, the 2" hose was able to flow 400 gpm. The trade-off for the lighter and more easily deployed hose was that a higher pump discharge pressure was needed. As previously mentioned, the 1¾" high-performance hose has higher friction loss than the 2½" hose. You may wonder if the higher pump discharge pressure is a big concern. In my opinion, it is not. First, it is important to understand that the higher pressure used on the

Table 10–4. Comparison of (a) 2½" handline with (b) 1¾" FDNY line with 5' section of 2½" hose at the nozzle, (c) 1¾" Combat Ready handline with a 5' section of 2½" hose at the nozzle, and (d) 2" Combat Ready handline with 2½" couplings

(a) 2½" handline

2½" Handline, 200'					
Nozzle Size	Flow Rate (gpm)	Nozzle Pressure (psi)	Nozzle Reaction (lbs reactive force)	Pump Discharge Pressure (psi)	Friction Loss (psi)
1⅛"	265	50	99	80	10
1³⁄₁₆"	300	50	111	85	20
1¼"	325	50	123	90	20
1⅜"	400	50	148	110	50
1½"	500	55	194	200	55

(b) FDNY 1¾" handline with 5' section of 2½" line at the nozzle

1¾" FDNY Handline, 200' length with 5' section of 2½" line at the nozzle					
Nozzle Size	Flow Rate (gpm)	Nozzle Pressure (NP)/Nozzle Base Pressure (NBP) (psi)	Nozzle Reaction (lbs reactive force)	Pump Discharge Pressure (psi)	Friction Loss (psi)
⅞"	160	50 NP	60	110	25
¹⁵⁄₁₆"	185	50 NP	69	120	30
1"	210	50 NP	79	145	40
1⅛"	265	50 NP	99	185	60
1³⁄₁₆"	300	50 NP	111	220	70
1¼"	325	50 NP	123	250	85
Metro	250	50 NBP	89	180	55

Note: The Metro nozzle is fixed gallonage nozzle manufactured by Task Force Tips.

hose is well within the designed pressure capabilities. The most common hose pressure ratings for attack line hose includes a 1,200 psi burst pressure and a 400 psi annual test pressure. According to NFPA and the manufacturers, the hose can be used up to 360 psi on the fireground.[12] This category includes the FDNY hose. The Combat Ready hose has a burst pressure of 1,500 psi, an annual test pressure of 500 psi, and a working pressure of 450 psi. We did not get anywhere close to the standard hose and high-performance hose pressure ratings on any

Table 10–4. (*Continued*) Comparison of (a) 2½" handline with (b) 1¾" FDNY line with 5' section of 2½" hose at the nozzle, (c) 1¾" Combat Ready handline with a 5' section of 2½" hose at the nozzle, and (d) 2" Combat Ready handline with 2½" couplings

(c) Combat Ready 1¾" handline with 5' section of 2½" line at the nozzle

1¾" Combat Ready Handline, 200" length with 5' section of 2½" line at the nozzle					
Nozzle Size	Flow Rate (gpm)	Nozzle Pressure (NP)/Nozzle Base Pressure (NBP) (psi)	Nozzle Reaction (lbs reactive force)	Pump Discharge Pressure (psi)	Friction Loss (psi)
7/8"	160	50 NP	60	110	25
15/16"	185	50 NP	69	120	30
1"	210	50 NP	79	130	35
1⅛"	265	50 NP	99	165	50
1 3/16"	300	50 NP	111	210	65
1¼"	325	50 NP	123	240	80
Metro	250	50 NBP	89	180	45

(d) Combat Ready 2" handline with 2½" couplings at the nozzle

2" Combat Ready Handline with 2½" Couplings at the Nozzle					
Nozzle Size	Flow Rate (gpm)	Nozzle Pressure (NP)/Nozzle Base Pressure (NBP) (psi)	Nozzle Reaction (lbs reactive force)	Pump Discharge Pressure (psi)	Friction Loss (psi)
1⅛"	250	50 NP	99	130	20
1 3/16"	300	50 NP	111	140	40
1¼"	325	50 NP	123	180	50
Metro	250	50 NBP	89	130	20

of the lines we tested. At 325 gpm, the highest pump discharge pressure obtained was 250 psi for the FDNY, 240 psi for the 1¾" Combat Ready, and 180 psi for the 2" Combat Ready with the 2½" couplings.[13]

Nevertheless, if the pump discharge pressure for either the 2" or the 1¾" Combat Ready is too high for your liking, you can try an alternative (fig. 10–10). On a 200' line preconnect, start off the first section of hose, which would be 50' of 2½" line. The friction loss in the 2½" hose at 50' with 300 gpm is 10 psi. Finish off the 200' of handline with 150' of Combat Ready 1¾" line, which will be more

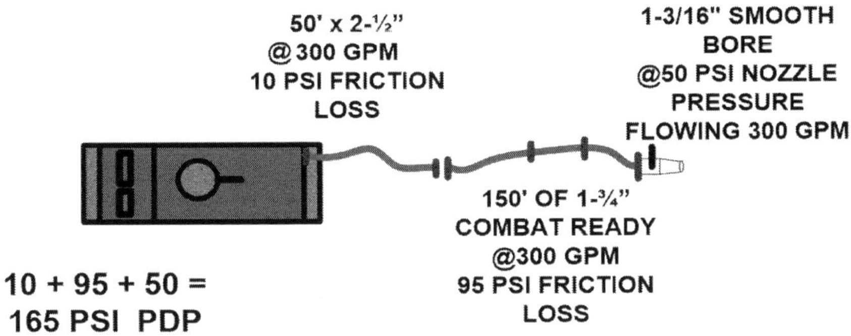

Figure 10–10. An example using Combat Ready hose

than enough line to bring interior for advancement. It has a total friction loss of 95 psi for the three sections that make up the last 150'. The friction loss is negligible on the 3' long, 2½" diameter stinger, which is placed on the end of the handline just before the nozzle to counter the nozzle whip. Thus we will not include it in these calculations. Add 50 psi for the 1$^{3}/_{16}$" smoothbore tip, and you get a pump discharge pressure of 155 psi. Compared to the pump discharge pressure for the straight 1¾" line, which is 210, this is a very respectable discharge pressure that is within the safe operations of a pump. Again, I have no problem with the higher pressures discussed previously, but the naysayers may find this pressure range more to their liking.

Table 10–5 shows the difference in weight between the hose sizes we have been discussing. A comparison of the weight of the lines offers a useful perspective on the weight savings possible.

The Design of the High-Rise Handline Based on the Facts: It's Pretty Simple

The use of high-rise handlines has been a frequent topic of discussion during my 30 years of teaching water delivery in the fire service. I have seen numerous changes, and most seem to work. Occasionally tactics and equipment seem to contradict other techniques, but in the final analysis, the fire gets extinguished. I have put together some sound handline water delivery operations, along with the proper equipment to get the job done. This section is based on information gathered from others, as well as what I have learned myself. Keep in mind that the contents of this section are concerned only with water delivery.[14]

Table 10–5. Hose capacities

Hose Capacities							
Hose Size	Volume of Water (gal/1')	Volume of Water (gal/50')	Volume of Water (gal/200')	Hose Weight (lbs/50')	Hose Weight (lbs/200')	Combined Water and Hose Weight (lbs/50')	Combined Water and Hose Weight (lbs/200')
1¾"	0.1249	6.24	24.98	20	80	72.10	288.4
1.88"	0.1442	7.210	28.84	20	80	80.20	320.8
2"	0.1632	8.16	32.64	24	96	92.13	368.5
2½"	0.255	12.75	51	25	100	131.5	525.9

Note: 0.7854 × diameter (in.)^2 × length (in.)/231in.^{3}/gal = volume in gallons. Gallons × 8.35 lbs/gallon = weight.

My goal is to provide information that will help firefighters decide what size handline to deploy in a high-rise fire. There are four basic components for a high-rise handline operation: the hose, the nozzle, a 2½" × 2½" gated wye, and an in-line pressure gauge.

Hose and nozzle

Choosing the proper hose and nozzle combination needs to be based on delivering the required flow for fire attack using low system operating pressures. The reason for this is simple. High-rise fire protection systems are limited in operating pressures because of elevation, friction loss in plumbing, pressure-reducing devices, and probably the most concerning issue, system pressures established by *NFPA 14, Standard for the Installation of Standpipe and Hose Systems.*[15] (See chapter 6 to review the two basic sets of NFPA standards, pre-1993 and post-1993.)

Because of the pressure issues in fire protection systems, most fire departments have opted to exclusively use 2½" handlines with smoothbore nozzles for all fire scenarios, big and small. The true fact is that the 2½" handline will provide the most water every time. Along with the positive flow capabilities of this line, however, there are also deployment and kinking challenges that must be addressed due to the size of the hose.

The low standpipe outlet pressures mentioned above can be misleading, causing firefighters to assume that smaller handlines such as 1¾" and 2" lines could not be used because of the higher friction losses of these lines compared to the 2½" line. It may be helpful to analyze the actual pressures that can be expected in a high-rise fire protection system.

Again, fire protection systems are based on a 500 gpm flow capability, and after flowing 500 gpm, the standpipe outlet residual pressure will need to be either 65 psi or 100 psi, depending on when the system was built. The true fact is that if 500 gpm is not flowing, which it probably would not be with the initial handline deployment, the standpipe outlet residual pressure will actually be higher.

The information from the Riviera Hotel flow tests, presented previously in table 10–2, supports this observation. The standpipe residual pressures increased significantly from the rated 65 psi pressure. Furthermore, all flows exceeded 150 gpm, which is well within the standard for interior attack lines. Keep in mind that the 65 psi system we tested these lines on is a worst-case scenario in regard to NFPA codes and restrictions. Any structure built after 1993 will have the 100 psi standpipe outlet pressure, which in turn means that the handlines will flow even more water. As a side note of interest, the Authority Having Jurisdiction in Clark County Nevada (Las Vegas) has increased the required residual pressure at the standpipe to 125 psi after flowing the required 500 gpm.

The size of the hose to be used should reflect the requirements of the fire and the system pressure. The water flow rate needs to match the heat energy generated by the fire. The diameter of the hose must be able to overcome any system low pressure issues.

Low-pressure nozzles are a must in a high-rise operation, even with the higher residual pressures that were indicated in table 10–2. Remember, system pressures will be low, so lowering the nozzle pressure from the standard 100 psi to a much lower pressure in the range of 40 psi to about 60 psi greatly improves water delivery capabilities.

It is important to understand that these flows were measured at the roof, which as mentioned is a worst-case scenario for pressure loss in the system. The lower you go in the building, the higher the flow and nozzle pressure will be.

Standpipe connections

Not all standpipe outlets are user-friendly in terms of connections to hose and gated wye appliances. Because of this, a short 15' section of either 2½" or 3" hose should be used to connect the high-rise pack to the standpipe outlet (fig. 10–11).

Figure 10–11. A 15' section of 2½" or 3" hose should be used to connect a high-rise pack to a standpipe outlet.

A 2½" × 2½" gated wye

A 2½" × 2½" gated wye should be connected to the end of the short section of hose leading from the standpipe to allow two handlines to be connected from one standpipe outlet. Flow tests at the same structure gave us a flow of 238 gpm each from two 2½" handlines 150' in length, with 1⅛". smoothbore tips.

A 2½" in-line pressure gauge

The 2½" in-line pressure gauge is connected between the gated wye and the short section of hose leading off the standpipe outlet (fig. 10–12). The purpose of the in-line pressure gauge is twofold. Part of the high-rise deployment procedure requires a flow test of the handline before commencing with the fire attack to determine if the required handline pressures can be met from the system. This in turn will help

> In order to have the luxury of an easily deployed 1¾" or 2" handline, it is imperative that a 2½" hose pack be brought into the high-rise structure in case it is needed.

Figure 10–12. A Task Force Tips in-line 2½" pressure gauge at the standpipe outlet, where the gauge has an adjustable angle at the elbow

decide what size line can be used. The other thing the in-line pressure gauge can do is help to dial in the required pressures for the handline. For example, say you are using Combat Ready 1¾" hose, and the required pressure for that line to flow 185 gpm with the 15⁄16" smoothbore tip is 105 psi. This can be noted with a label on the side of the gauge body so if an adjustment is needed at the standpipe, a firefighter simply adjusts the standpipe outlet valve to the required pressure.

11

THE BIG HIT OFFENSIVE CONCEPT

Introduction

I have wanted to write about the big hit offensive concept for a long time now. My goal was to keep it basic while still conveying as much useful information as possible. I discussed the information with my wife and realized I could easily illustrate the concept though a comparison to camping. When we arrive at a campsite in our RV, we often set up a small campfire. Sometimes the fire is still burning when we are ready to call it a night, in which case we like to put the campfire out instead of letting it burn unattended. I asked my wife, "If you had a choice between putting out the campfire with a squirt gun or a garden hose, which would you choose, and why?" Her answer was simple. She replied, "I would choose the garden hose because it would put out the fire much faster."[1]

I enjoy watching videos of large fires not only for entertainment, but also for educational purposes. Once I asked my wife to watch one of these videos with me. Since I was not present at this fire, I do not consider myself in a position to conduct a critique of the fire department's operations. That would not be fair. Instead, I want to use it as a learning tool. This video showed a three-story, wood-frame residential complex, approximately 50' × 100', with the entire front of the structure well involved. The first line out on this large fire was a 1¾" handline, which went to the exposure. The second line out was a 2½" handline outfitted with a triple-stack smoothbore tip combination, with the entire stack still attached to it. In other words, they were using the 1" tip, and if it was pumped correctly, they were getting a flow rate of 210 gpm.

My wife and I have been married 35 years now, and I have been teaching on this topic for the same amount of time. She has thus become semi-versed with my concepts and theories. As we watched this video, I asked her if there was anything else that they could be doing that might put this fire out faster and save the exposure. She replied without any hesitation, "Why don't they fire up the deck gun and blast the <expletive> out of it?"

The concept of using large flow streams to put out fires is actually pretty simple, and as far as firefighting goes, it is also fairly safe for the firefighters. Have you heard the statement, "You have to match the gpms to the Btus"? I go one step further and say, "You have to overwhelm the Btus with the gpms" (fig. 11–1).

Here is a brief list of reasons why some fire departments do not practice the overwhelming blast concept:

1. Firefighters may not understand their equipment and capabilities.
2. Firefighters may not understand the fire science of overwhelming the generated heat energy with a sufficient water flow rate.
3. Firefighters may lack training with this concept.
4. Firefighters may not fully understand water delivery, not just from the discharge side of the pump but also in regard to water supply.
5. Firefighters tend to do what they have always done.
6. It's not "fun."

Figure 11–1. This engine is moving a ton of water in a simulated high fire load. Notice it's being supported by a relay pump operation.

If your department lacks sufficient water delivery for large fires, you can probably relate to a few of these reasons, and possibly others that I did not mention. The bottom line is that we need to start practicing big hits on certain fires to do a better job of controlling and eventually extinguishing the fire.

My theory on exposure protection is that if there is an exposure problem with first-in companies, I need to make a decision. Do I protect the exposures with water application, or do I knock down the problem that is creating the exposure, i.e., the fire? If you think that a direct attack on the fire can knock down the fire in seconds to a point where it is not hot enough to create the exposure hazard, then go for it. If not, the direct application of water on the exposure is warranted. If you choose to hit the source of the exposure threat, then knock the snot out of it. Hit it with everything you possibly can.

Two topics from my above-mentioned list will greatly improve the knockdown capabilities at large fires. When I discuss water delivery, I usually start with the water supply. In this section, however, I want to begin with required streams and then talk about developing the required water supply to feed the streams. The reason for this is simple. Often bigger tips or larger flows are not used simply because the water supply is considered insufficient.

Deploying a 500 gpm Blitz Attack from Tank Water

This section will focus on fire attack on an exposure-threatening fire based on a limited water supply, more commonly known as a *blitz attack operation*. It will be a worst-case scenario with a 500-gallon booster tank and will use 500 gpm for the flow rate for maximum knockdown power.[2]

The most common initial hose deployment that departments use when they are facing a threatening fire with only tank water is to pull a small handline, more than likely a 1¾" line, which is used to protect exposures. The flow will usually range from 100 gpm to 150 gpm. The reasoning behind this choice, besides protecting exposures, is to conserve water until a water supply can be established. At 150 gpm, a 500-gallon booster tank will last a little over three minutes. This might be enough time to access the uninterrupted water supply if the first-in company laid the supply line, but it will probably be a close call at best. While many departments go straight in on tank water and call for the second due company to lay a line, they are really pushing their luck. What if the second due company is delayed or does not bring in the line? Now they are really in trouble. The fire that has caused the exposure problem is still causing the problem, and there is no more water.

Yes, water puts out fire. The rate of extinguishment is based on the flow rate (gpm) of the water delivered onto the fire. Throughout the years, articles have been published discussing statistics in regard to water and its abilities to put out fire, in hopes of improving the process. They have used terms such as *Btus, rate of application, fire growth rate, big drops, little drops*, and so on, but what do they really teach us about the application of the water? The reality is simple: put enough water on the fire to put it out as quickly as possible. When we make our attack, we do not think about all of the more technical aspects. Our goal is to apply water at a high enough flow rate to extinguish the fire as quickly as possible, or to at least slow it down.

With that being said, I would like to offer my perspective on what needs to be done to have a successful outcome. If the company officer thinks it is possible to extinguish the fire with a quick blast, do not mess around with putting water on the exposure. Instead, just blast the fire! Put enough water on the fire to achieve at least a knockdown as quickly as possible. A good definition of a knockdown on a fire for the blitz attack scenario is to hit the fire with an overwhelming amount of water to change the state of the fire from fierce, out-of-control, and spreading to a more docile state that does not threaten the exposures. This will allow firefighters to regroup and get the proper lines in place to accomplish extinguishment. It is much like a one-two punch. This tactic, if sized up and conducted correctly, will have an excellent chance to achieve a knockdown on a fire in seconds. This means the exposure will only be exposed for seconds instead of minutes. It is entirely possible to achieve a 5-second knockdown with the correct weapon and ammunition (flow rate) of choice. This holds true from the smallest of fires to the largest, wherever an initial attack can be successfully carried out.

With a limited water supply, the application time for a blitz attack should be no more than 30 seconds. This goal, as challenging as it may sound, has a good chance of being accomplished with the right flow rate. Always try to flow a maximum amount, even if you think it is overkill. If you manage to flow more water than the fire requires, the only thing that will happen is that the fire will go out more quickly.

When it comes to proper water delivery tactics, there are several factors to consider based on the situation at hand. These include the water supply, amount of water needed (flow rate), water delivery system, manpower, and hose-handling techniques, if applicable.

No matter which method of attack will be used, you need the water to accomplish it. Engine company booster tank operations need to be more precise in this type of operation because there will be only one chance to extinguish the fire quickly. One big question arises when the fire is creating exposures: With a limited water supply, do we protect exposures or conduct a fire attack? The best way to protect an exposure is to eliminate the source of the exposure hazard, or in other words, the fire. A thorough size-up can determine whether or not hitting the fire first will accomplish an immediate knockdown. This is where real-world experience with water volume versus fire volume (gpm vs. Btus) comes into play. With that being said, what if the company officer does not yet have that experience?

The National Fire Academy (NFA) has developed a formula [length (ft) × width (ft) /3 (ft^2/gpm)] that accurately calculates the required flow in gallons per minute for a structure based on dimensions and fire involvement.[3] Can the formula be used at the time of the fire? Obviously, it is possible. It will be difficult at best, however, due to the extreme situation at hand. Nevertheless, this formula can be used as a training tool to determine the types and sizes of structures in your area that could be candidates for 500 gpm blitz attack response.

Water supply

The first thing to know with this type of operation is how much water is available and at what flow rate (gpm). Most plumbing designs for connecting the booster tank to the pump only allow a maximum of 500 gpm to be delivered from the tank to the pump, so many departments adopt this as a minimum standard when designing their engines. So, should 500 gpm be used as the flow rate? Obviously the company officer will make the choice based on the fire volume. For now, let's say that 500 gpm will be the required flow.

If you are going to attempt a 500 gpm blitz attack from a 500-gallon booster tank, many factors must come together. First you must make sure the fire is a 500 gpm fire or less. Again, this is based on being able to get a knockdown within 30 seconds. These fires could include garage fires, mobile home fires, fully involved houses under 2,000 ft^2, commercial properties approximately the size of a convenience store, and so on. Remember, you are only trying for a knockdown.

It is important to realize that flowing 500 gpm does not mean you will use 500 gallons of water. This is merely the rate of flow. Think of the flow rate as it relates to using up the water supply in units of gallons per second, since the knockdown needs to be accomplished in 30 seconds or less. A flow rate of 500 gpm is 8.3 gallons per second. The following sequence of photos (fig. 11–2) shows a well-involved two-story residential structure that was hit with a 500 gpm stream that got a knockdown in 16 seconds. The total amount of water used was 132 gallons from a 500-gallon booster tank.

Deck gun

The method of delivery depends on the manpower available to put it into play. There are two methods for implementing the big hit. The first is the fixed master stream, also known as the *deck gun*. This is a one-person job whether it is a manually operated appliance that requires the firefighter to be on top of the apparatus or is remotely operated from the ground. It is important to know the exact pump

Figure 11–2 (a). Demonstration of a 16-second knockdown with the Big Paulie 500 gpm nozzle

discharge pressure for the deck gun in order to avoid cavitation of the pump by overpumping the device. In reality, even though the flow rate of 500 gpm is standard, most apparatus can deliver a little more. Nevertheless, the 500 gpm target

Figure 11–2 (b). Continued

flow should be kept. It is also important to avoid wasting water in the application of the fire stream. Having the deck gun pump discharge pressure obtained before the appliance is opened will help.

Figure 11–2 (c). Continued

One problem that could arise from doing this is opening the discharge under the required pressure if the discharge mechanism is the type of handle that works by pulling the rod out. NFPA requires a slow-moving device to open and close the master stream valve, which involves a wheel/gear type of mechanism.[4] This will alleviate the problem.

Figure 11–2 (d). Continued

The deck gun should be aimed at the target as much as possible, again to help prevent wasting water. Finally, use only as much water as it takes to achieve a knockdown, not an extinguishment. When an uninterrupted water supply is secured, the gun can be reopened to complete its job.

Nozzles

There are two types of nozzles that can be used in a deck gun blitz attack: a combination nozzle (automatics are the most common) and a smoothbore tip. Velocity and penetration can be crucial in an initial blitz attack to hit as deep into the seat of the fire as possible. This means that a high nozzle pressure should be used. Combination nozzles are usually rated at 100 psi but can go as high as 120 psi. In this case, the higher the better. The 1⅜" smoothbore tip is rated to flow 500 gpm at 80 psi nozzle pressure. Using smaller size tips to get 500 gpm can also be accomplished without breaking the rules set by the manufacturers. For example, the 1¼" tip flowing 500 gpm has a nozzle pressure of 115 psi. The 1⅛" tip can also flow 500 gpm at a whopping 175 psi nozzle pressure.

Going back to the 30-second rule, if after flowing the chosen stream it is decided that a knockdown does not seem possible, do not use any more water. Shut it down and regroup.

Handlines

Now let's talk about handlines. More than likely, if a high-flow handline is to be used for a blitz attack, the firefighters will be in a stationary position. If this is the case and the situation will allow, do not stand up with the line, especially at the higher flows. It will beat you up and may indirectly reduce the flow rate if the firefighter at the nozzle needs to gate it down to make it easier to handle. Instead, just have a seat. It is a proven fact that the firefighter's weight sitting on the hose is extremely helpful in eliminating the nozzle reaction effects. If a 500 gpm attack line is to be used, it will take the combined weight of two firefighters to hold down the nozzle reaction effects. If lower flows are delivered, it may only require one. Training in whatever line you choose will help you decide what works best.

The 500 gpm line will need to be a 2½" line. Flow tests have proven that a 2½" line can be up to 200' in length and provide a 500 gpm flow at around 200 psi pump discharge pressure (fig. 11–3). Of course, the design of the discharge plumbing will dictate the actual pressure needed. The 1⅜" tip at 80 psi nozzle pressure, a 1½" tip at 55 psi nozzle pressure, and a 500 gpm combo nozzle at 80 psi or 100 psi nozzle pressure are all good nozzle combinations.

This same 500 gpm application can also be delivered through what I like to call the "mini monitors." Basically, the mini is a small version of a portable master stream, light in weight and capable of a 500 gpm flow (fig. 11–4). It has a single inlet and can be supplied from a single 2½" line.

Every apparatus should have at least one designated handline that can be used for a large flow operation, whether it is preconnected or simply stored in the hose bed. It should have a nozzle that will flow the maximum amount of water based on the equipment and policies of the department. A good example would be a preconnected 2½" handline 200' in length, with triple-stack smoothbore tips that are all connected to the nozzle. I personally do not see the advantage of using a triple-stack tip because even though it gives a firefighter the opportunity to dial in the flow, this never seems to happen. The entire stack is almost always left on the nozzle, which means that the 1" tip is what is flowing. If it is pumped correctly, the flow is 210 gpm. Say that the firefighter on the nozzle wants to go bigger and go to the 1¼" tip to get 328 gpm. In order to do this, a different pump discharge pressure must be used. Now it becomes a communication issue because the pump operator needs to know what size tip is being used on the nozzle to get the correct flow. Often this can be a problem. My recommendation is to choose one size

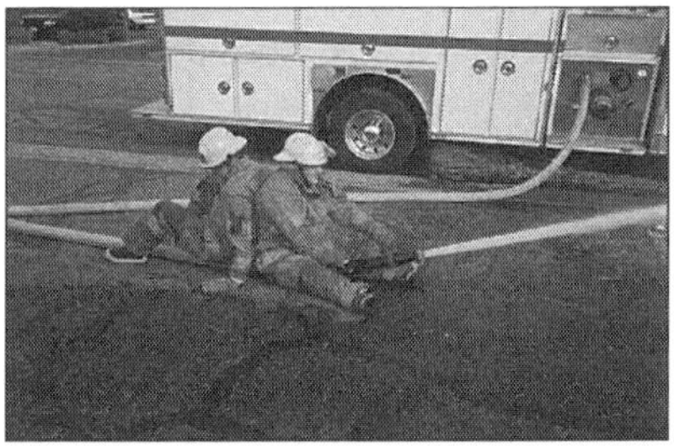

Figure 11–3. A 2½" line that is 200' in length can provide 500 gpm at around 200 psi PDP.

Figure 11–4. Flow rate of 500 gpm delivered from the minis

nozzle in regard to its flow capabilities and nothing else. If the preference is smoothbore tips, choose the one tip that you would like to get the biggest flow from and keep it on there. Instead of keeping the triple-stack tip on, just have the 1¼" tip. With this setup the firefighter will know what to expect, and the pump operator will know what to supply.

If you choose to go with a combination nozzle, whether it is automatic or fixed gallonage, I suggest going to the maximum flow available from the nozzle. Most combination nozzles have a maximum flow ranging from 250 to 300 gpm. The key is to have one flow so there is no confusion about what will be delivered on the handline. If it is an automatic nozzle, my suggestion would be to pump to the maximum flow capacity of the nozzle. For example, a 100 gpm to 300 gpm automatic nozzle should be pumped as though it is flowing 300 gpm. If the firefighter needs to gate down the nozzle, that can be accomplished without communication to the pump operator.

Master Streams

Now it is time to increase the size of the fire. Previously I mentioned a video on YouTube showing a fire at a three-story apartment complex that was well involved in the front. Now the response changes since we are talking about multiple streams of at least 1,000 gpm. Again, a study of YouTube fires that are multicompany operations with the big guns flowing reveals that often the entire stack of tips is left on the master stream appliance throughout the duration of the fire. In fact, in one video I could hear the order being given to use the 1⅜" tip. If that tip is being pumped properly, a flow rate of 500 gpm should be expected. The problem is that the fire in this instance was probably a 2,000 gpm fire (fig. 11–5).[5]

The same decisions must be made concerning whether to protect exposures or hit the fire. This time, however, an additional important and possibly more complicated decision also must be made if you are going to hit the fire with the proper flow. Where will the water be supplied from? The fire on YouTube was in a major city and involved a major fire department, meaning that apparatus and fire hydrants were likely plentiful. This department used 5" large diameter hose and grabbed the closest hydrants to the fire. I am not sure why they used the 1⅜" tip, but a common reason for this could be because they ran out of water even using large diameter hose. I have a phrase that I like concerning this situation: "The water is out there. You just have to go get it." Most departments with large diameter hose lay from the closest hydrants, and when they develop water supply issues, they give up. Word gets back to the incident commander (IC) that they ran out of water, and the IC tells everybody to gate down to the

Figure 11–5. A 2,000 gpm platform being supplied by an engine

point where they can get streams that will reach the fire. Eventually the fire is extinguished.

Here is what needs to be done. First, if a master stream is going to be used for a sustained water delivery operation, or in other words not a tank water blitz attack, the goal should be to supply the rating of the appliance. Most fixed mounted master stream appliances on engine companies are rated to flow 1,250 gpm. This is not achievable with the 2" tip, which is the largest in the stack because of appliance restrictions. A 1,250 gpm master stream appliance has a maximum allowed inlet pressure to the device of 200 psi. Based on the 1,250 gpm flow with a 100 psi nozzle pressure, a combination nozzle can only flow up to 631 lbs nozzle reaction. In my testing I have found that a 1,250 gpm appliance cannot reach 1,250 gpm with the 2" tip because of the nozzle reaction limitations. With that being said, either use a combination nozzle for 1,250 gpm or set the goal for a fixed master stream with the 2" smoothbore tip for 1,000 gpm.

Here is another example of a maximum flow operation from a single unit. The unit is a 2,000 gpm elevated platform with a pump. Departmental procedures require a 2,000 gpm flow, and there is an evolution that will produce it efficiently. The quint gets the initial water supply with 5" hose from the closest hydrant. Unless the fireground area is already being tapped by multiple hydrant use, or if it is known that the hydrant is weak, I do not see a need to make this line a relay pump operation. It would not be wrong to set it up as one, however. Most hydrants cannot supply 2,000 gpm, but even if a hydrant can, one of the issues that will be encountered with this operation is an extremely high engine speed from the engine on the quint. I recommend automatically laying a second 5" supply line back to a hydrant and setting up for a relay/tandem pump operation. This will bring in extra water if needed and will assist the engine of the quint by lowering the demands on it by means of supplying additional pressure from the relay (fig. 11–6).

Whether a high-flow handline or a master stream is going to be used, it must hit the fire. By this I mean that you should move the stream around to hit as much fire as you can from your location. So often I see that when a big stream is aimed into a fire, it looks like a big drill trying to drill a hole. The stream does not move. It almost seems like they are trying to fill the building up with water from that point.

Water Supply Evolutions

When a multiple master stream operation is going to be utilized to fight a sustained fire, obviously there must be a water supply to deliver this flow. If the water is not available from the closest hydrants from the initial supply line, another engine needs to reverse out to a more distant hydrant that is hopefully on another

Figure 11–6. These streams can be doubled with the right water supply.

loop and begin relay pumping. These operations were common before the advent of large diameter hose. "Big water" meant putting pumps on hydrants to pump to the engines that were delivering the water. Large diameter hose has changed water delivery in many ways. Realistically, however, the real magic of large diameter hose is its use for small fires, which allows hydrant pressure to deliver the required flow. When it comes to the big fires where multiple master streams are needed, more than likely we need to go back to the old days in regard to our operations and set up a relay pump operation to move the required water. The good thing is that large diameter hose used in the relay really will move a significant amount of water, especially compared to the dual 2½" or 3" lines used in the past. As a side note, it takes five 2½" lines to equal one 5" line, and four 3" lines to equal one 5" line (fig. 11–7).[6]

The IC should not wait for water supply issues to arise but instead should expect them and immediately start the resources in motion to set up for large water supply evolutions. The IC should designate companies to start the reverses to other hydrants and set up for relay pumping. If the extra water is uncertain at the time, the supply lines should not be charged. Only the hookups to the hydrant should be charged. If it turns out that the water is not needed, an uncharged line will be much easier to pick up. If the line is already laid, it will only take a couple of minutes to charge it. On the other hand, if you wait to lay the line until the

Figure 11–7. Large diameter hose such as 5" line works best for supply lines.

water is needed, the time frame for water is likely at least 15 to 20 minutes. It is better to be proactive (fig. 11–8).

Because of the complexity of the concepts I have discussed and how they can be implemented, especially on large fires with multiple master streams, freelancing is not an option for implementing the operation. Can you imagine individual units trying to set up multicompany operations that require specific hydrants and specific hose lays? A water supply or water management officer or officers must be incorporated into the command structure specifically to design the hose evolutions. The water supply officer takes orders from the incident commander regarding the required flows and appliances to be used and then implements the plan with the required hose and apparatus.

Figure 11–8. Maximizing all the ports on the hydrant with large diameter hose

Attack Units

Regardless of the method of attack, engine company booster tank ope should be precise, especially if you only have one attempt to knock dc fire (fig. 11–9).

First alarm

The first alarm consists of E-1, E-2, E-3, E-4, and T-1 (figs. 11–10 through

Second alarm

The second alarm consists of E-5, E-6, E-7, E-8, and T-2 (figs. 11–14 t 11–17).

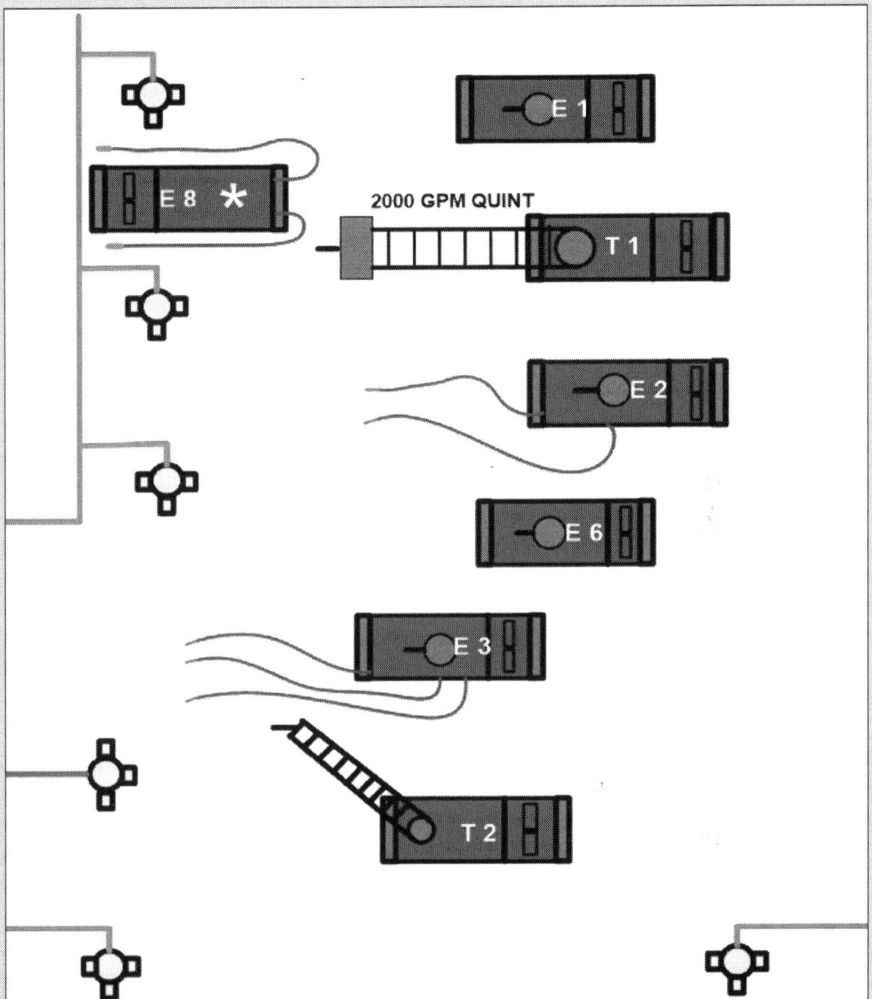

Figure 11–9. You need water no matter what method of attack will be used. Engine company booster tank operations should be precise.

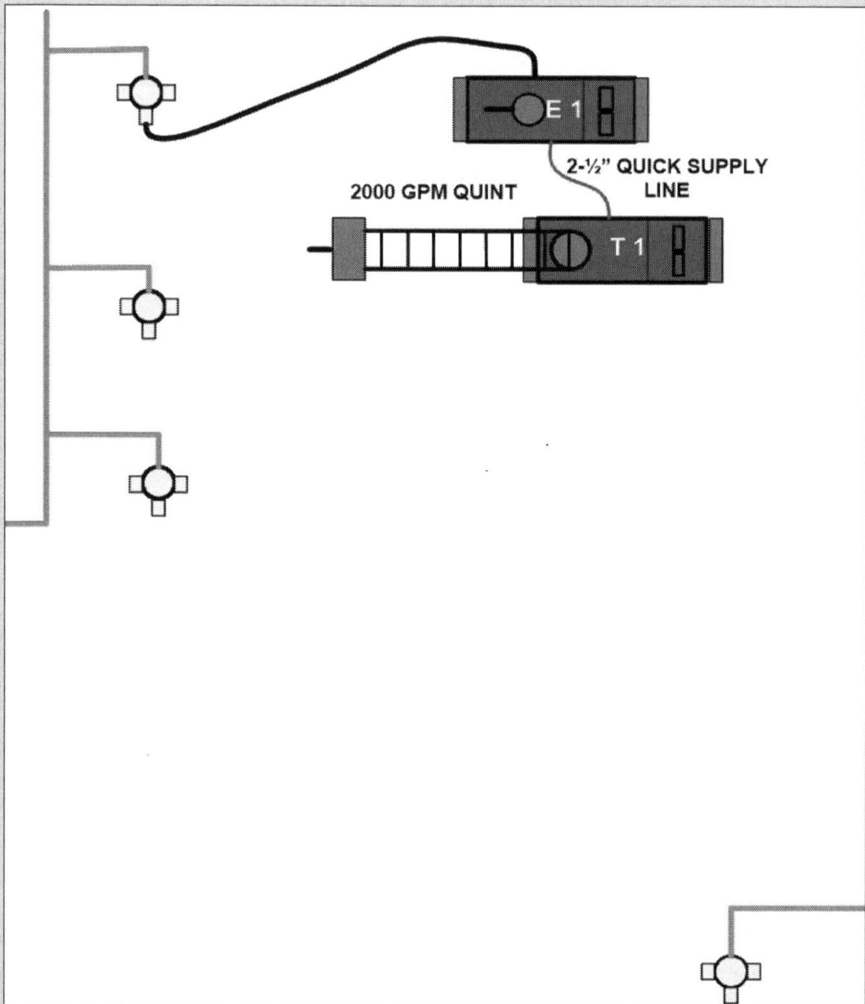

Figure 11–10. Start a blitz attack with tank water while you wait for your water supply. If you do not get a knockdown by the time you get a water supply, change out to the 2" tip and send any extra water to T-1.

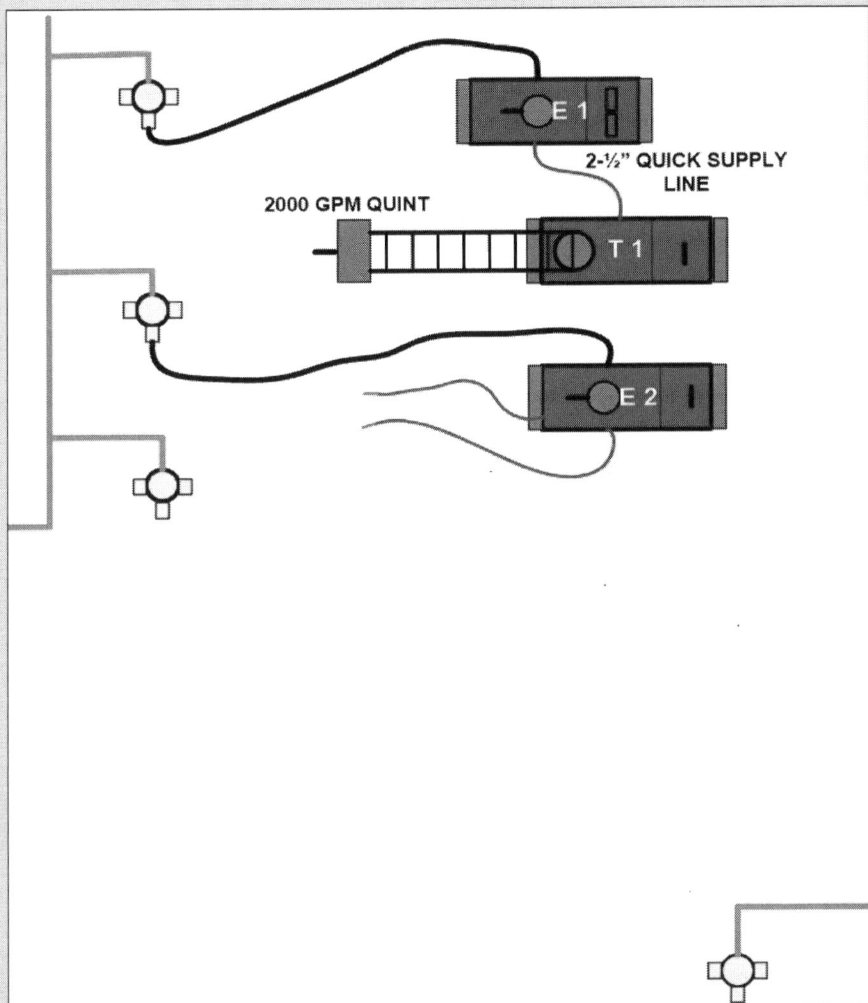

Figure 11–11. E-2 should lay a supply line and deploy 2½" handlines.

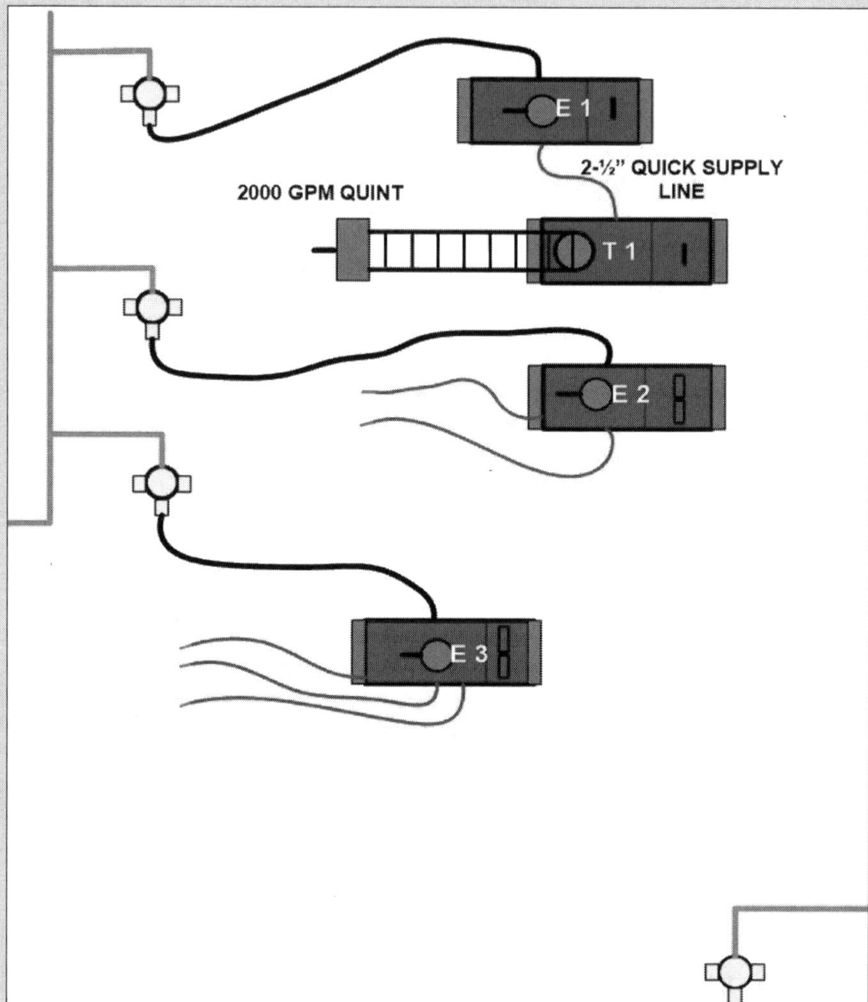

Figure 11–12. E-3 should lay a supply line and deploy 2½" handlines.

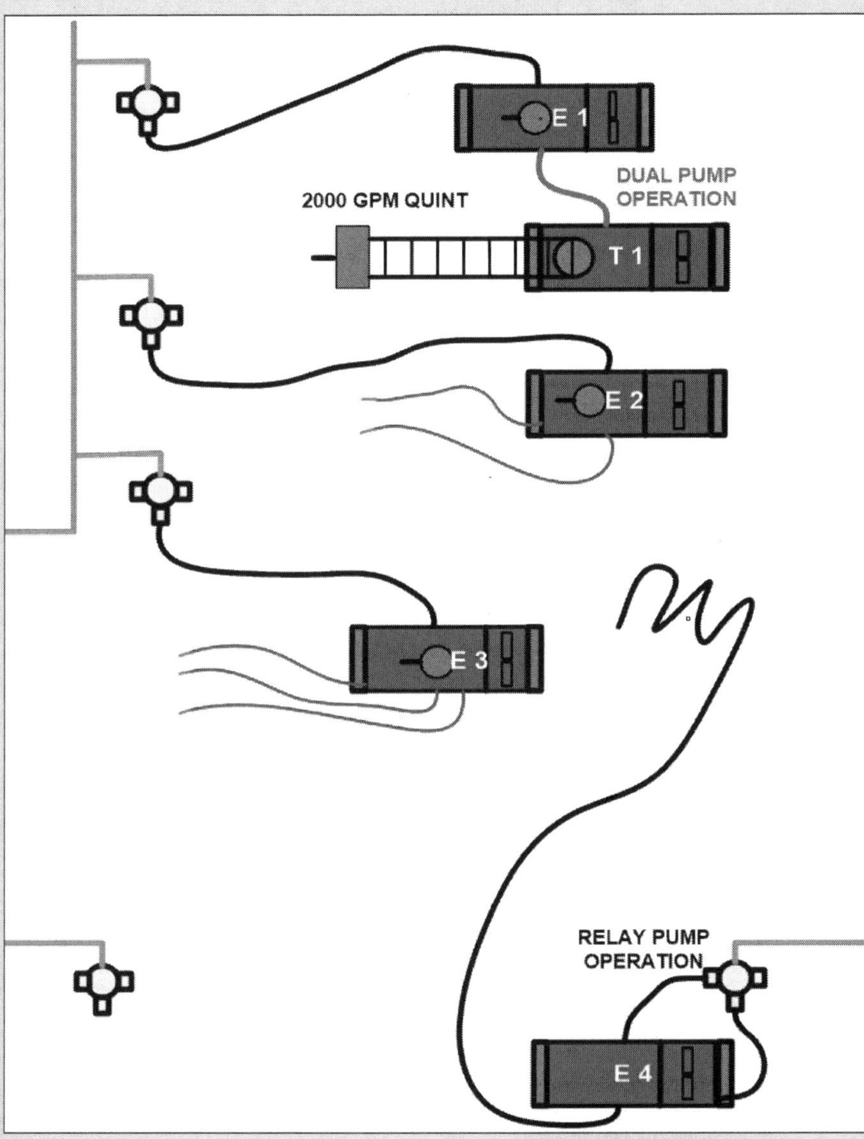

Figure 11–13. E-4 should reverse out from the area of E-2 and T-1, lay out to a hydrant, and set up for relay pumping. They should not charge the relay line at this point. It will not be connected until it is known where it will be needed. E-1 should set up a dual pump operation with T-1.

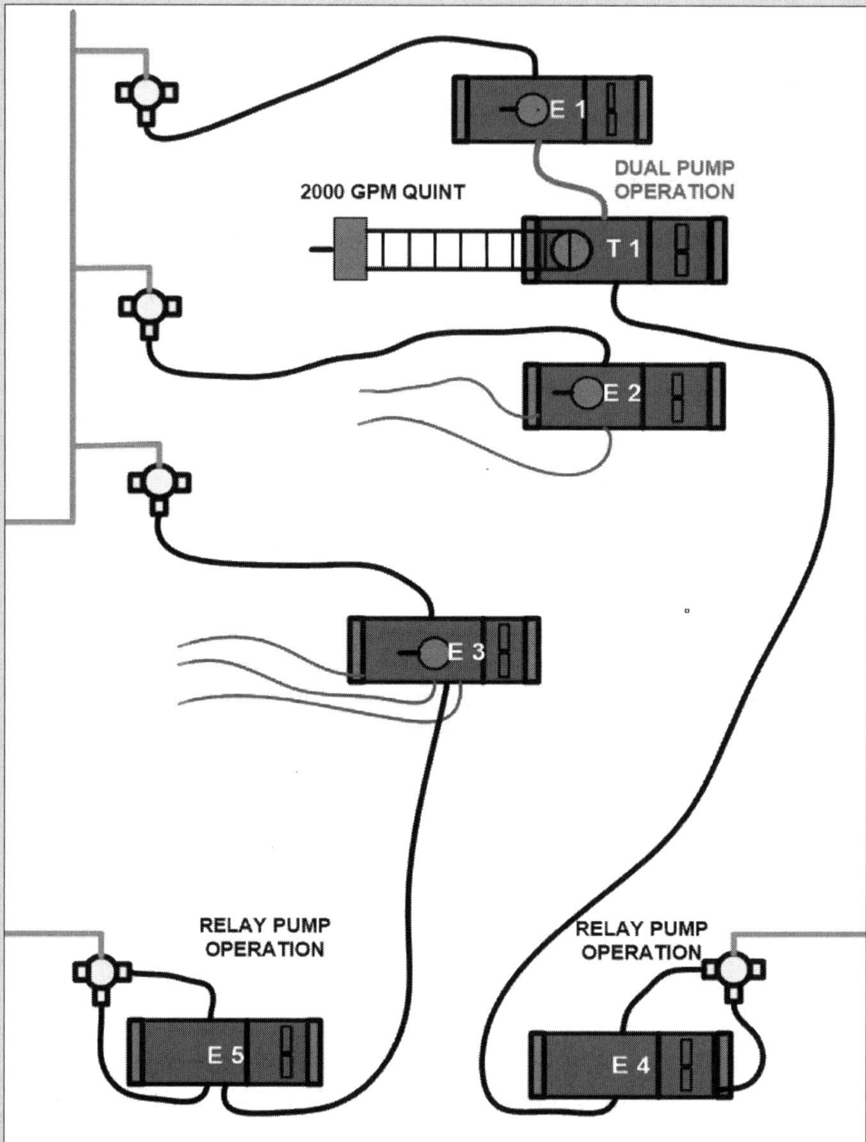

Figure 11–14. E-4 should now be connected to T-1 and start the relay with them. E-5 should set up a relay with E-3.

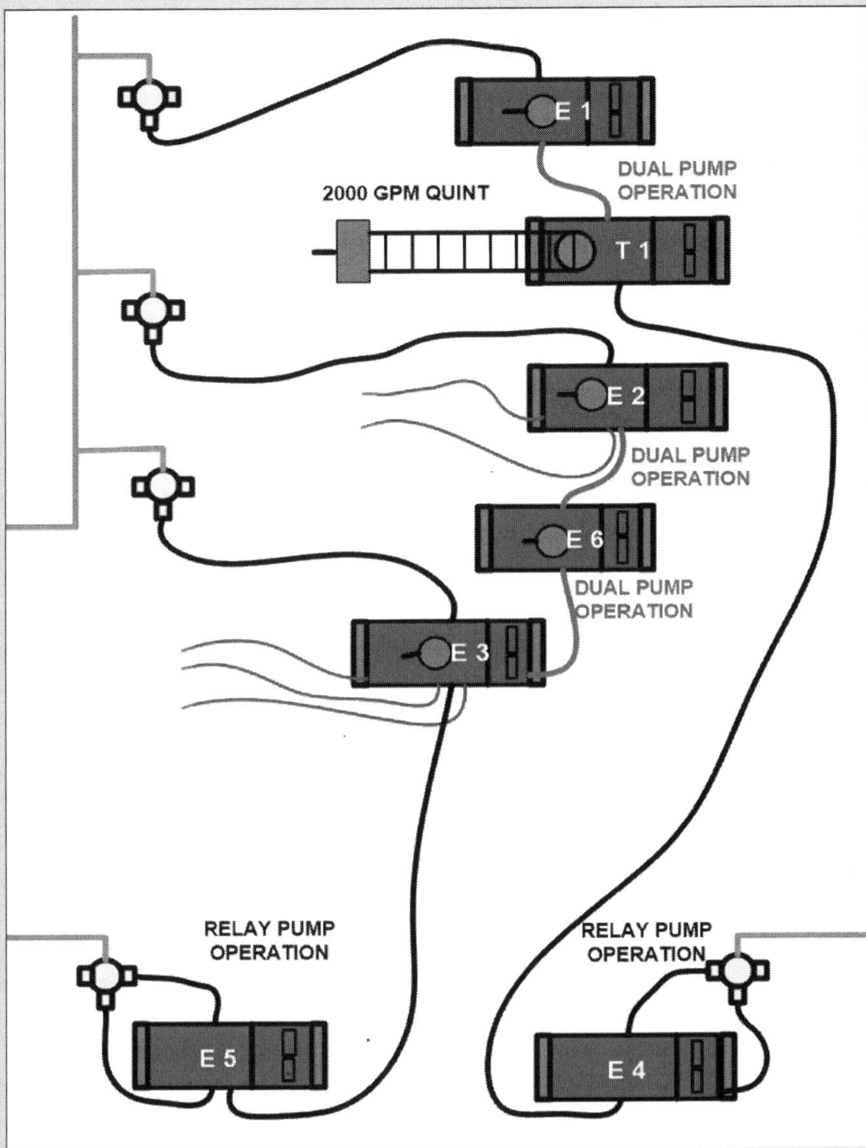

Figure 11–15. E-6 should spot their engine between E-3 and E-2 and set up dual pump operations between them.

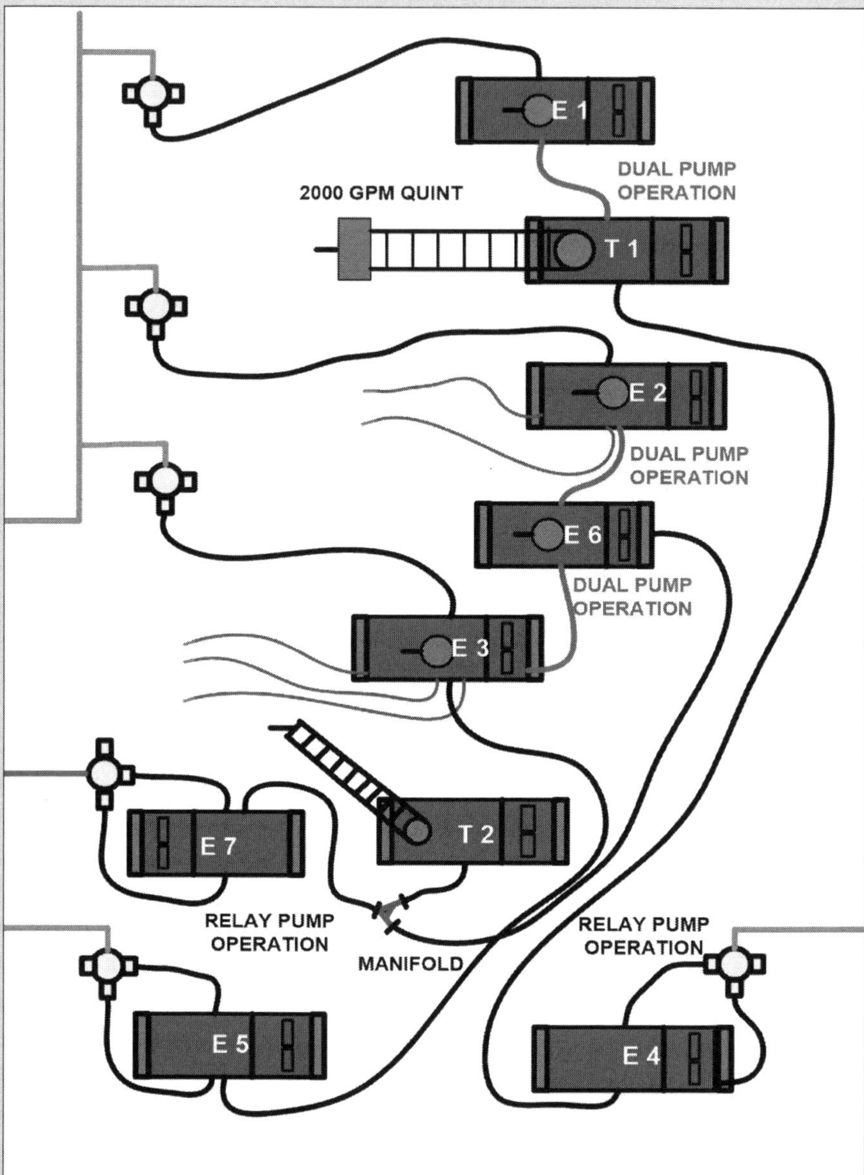

Figure 11–16. T-2 should set up for an elevated stream. E-6 should connect a supply line to the manifold to supplement their water supply operation. They will be getting a water supply from E-7 through a relay via a manifold at their location.

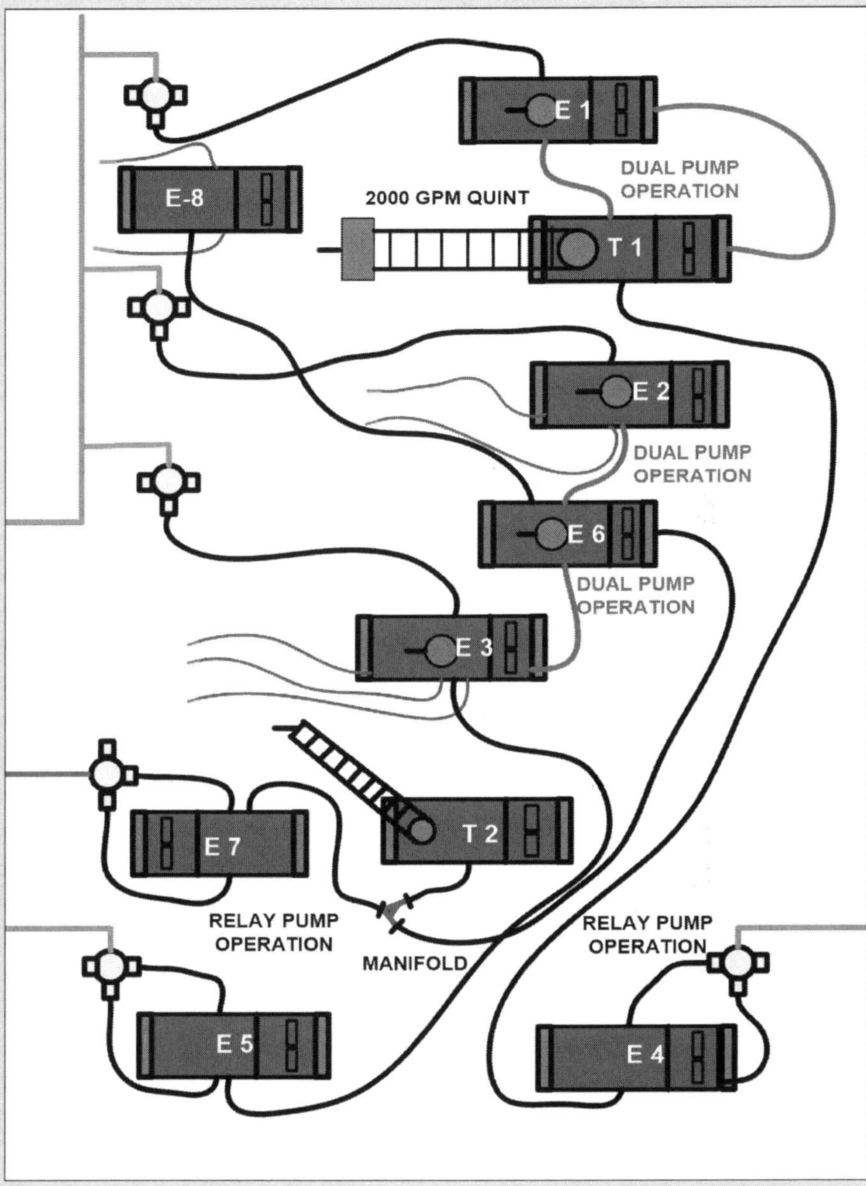

Figure 11–17. E-8 should deploy 2½" handlines just in front of T-1 and get a water supply from E-6. T-1 should tie into E-1 with a second dual pump line.

"This Is Great Stuff, but We Only Have Two Stations"

Many departments are not capable of some of the large flow water delivery evolutions presented here simply because they do not have the apparatus and manpower. In some cases, it may be possible for departments with fewer resources to access mutual aid responses. If mutual aid resources are available, even if they are not right next door, it is extremely important to figure out a way to use them efficiently. There are several ways this can be done. One crucial adjustment that can be made is to have an automatic dispatch that sends mutual aid units at the very moment the first-in units are dispatched. Another step that can be taken is to form strike teams for water supply delivery only. Upon arrival, these units immediately being to stretch supplemental supply lines to the units that are already at work.[7]

What about the departments with only one or two stations that unfortunately have no help coming? If this is the case, many tactical decisions must be made concerning how to deploy whatever manpower and equipment they have for the emergency at hand. For example, a decision may need to be made that stops all interior attacks and just deploys heavy streams to attempt to save the block. A good example of this was used in the Los Angeles riots in 1992 when hundreds of major structure fires were burning within the city at the same time. Even though the fire departments ran a statewide mutual aid response, many large commercial fires were handled by a couple of engines.

The bottom line is that where there is a will, there is a way. It may not always turn out to be the best operation possible, but you would be surprised what a well-tuned group of firefighters can do when they get in a pinch. It does require thorough training.

12

MEASURING WATER FLOW FOR FIRE SUPPRESSION

Introduction

Since water is the primary extinguishing agent for fire suppression, it is important to be able to deliver the required flow rate at the required pressure. There are a number of factors that can negatively impact water flow in water delivery systems on the fireground (fig. 12–1). Water delivery systems, also called hose evolutions, must take into consideration elevation gain or loss, nozzle pressures, and friction loss in hose, appliances, and plumbing, all of which can impede water flow. Firefighters have the ability to overcome pressure restrictions within the specifications of the equipment being used, mainly by increasing the pump discharge pressure on the engine. Hose evolutions can also be designed to overcome some of the pressure restrictions.[1]

Firefighters must be well trained with all of the equipment and procedures involved in their water delivery deployment. This knowledge should concern everything involved with moving water according to the operational guidelines of their department. This is something that cannot be accomplished at a fire without first properly conducting flow tests.

The best way to establish the water flow statistics to be used in designing hose evolutions is to conduct flow tests. Flow tests give you specifics on the flow rates and pressure requirements for different evolutions and for equipment such as hose, nozzles, and appliances. This section will discuss specific tests that can be conducted and will review the necessary equipment and procedures.

My background in teaching water delivery has involved nearly every flow test possible over the last 35 years. Like most of you, I have learned a great deal from the school of hard knocks. I believe it is extremely important to contact the manufacturers of the flow testing equipment to learn how it is best used.

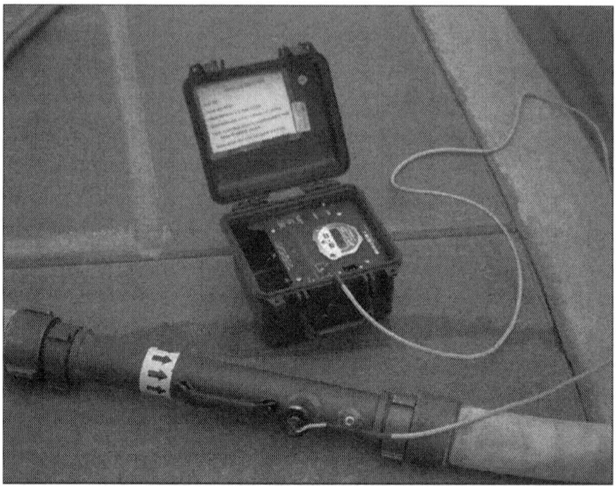

Figure 12–1. A flow test specifies the flow volume and pressure requirements for different evolutions and equipment such as hose nozzles and appliances.

Photo courtesy of Paul Shapiro.

Pressure Gauge/Pressure Measuring Equipment

Pressure gauges have been used in the fire service for many years for measuring water flow. There are two types of pressure gauge instruments that are used for flow testing. One is an in-line pressure gauge and the other is a pitot gauge (fig. 12–2), as discussed in chapter 4. The in-line gauge is considered a *wall gauge* (fig. 12–3). In other words, the gauge probe is inserted into the gauge housing from the sidewall. The pitot gauge measures the nozzle pressure of a fire stream, which is then referenced on a chart for the correct flow rate. It has a blade with a very small hole on the end that is inserted into the fire stream and sends the pressure to the gauge for a reading. The pitot gauge can either be handheld or set up in a stationary fixture.[2]

The pressure gauges used in water delivery flow testing in the fire service typically have 2 psi increments or 5 psi increments. The 5 psi increment gauge is the most commonly encountered. Some say that the problem with the 5 psi increment gauge is that you have to guess what the exact pressure reading is between the increments. More than likely this will not be crucial in your flow tests, however.

I have found that when I measure nozzle pressure with a pitot gauge during flow tests, the rise in pressure from the engine pump does not correlate perfectly with the resulting increase in the nozzle pressure. For instance, say that the test required the pump operator to raise the discharge pressure 10 psi. The nozzle pressure will probably not experience a 10 psi increase because generally when

Figure 12–2. The pitot gauge measures the nozzle pressure of a fire stream, which is then referenced on a chart for the correct flow rate.

Figure 12–3. An in-line gauge unit from the sidewall

the pressure is raised, so is the flow rate, potentially producing greater friction loss. The higher friction loss eats up some of the additional pressure, and so the nozzle pressure does not increase evenly with the discharge on the pump. Figure 12–4 shows a simple test that verifies this theory.

I prefer to round off the pressure reading to the next highest increment (fig. 12–5). Flow tests have proven that doing so results in little or no increase in flow rate and thus does not appreciably affect the flow test. Rounding off the pressure reading on the gauge makes it simple to read on the fireground.

All pressure gauges should be calibrated at least annually, but preferably three times per year if you are conducting numerous flow tests. When more than one gauge is being used for the flow tests, as would be the case for calculating friction loss in fire hose, it is important to make sure that the two gauges read exactly the same. A good way to test for gauge accuracy is to put the gauges together connecting one side to a discharge of the pump or a fire hydrant and capping off

Friction Loss Effect on Flow	
Tip:	1⅛-inch
Nozzle Pressure:	50 psi with a flow of 265 gpm
Pump Discharge Pressure Increase:	5 psi
Actual Nozzle Pressure Increase:	No gain in nozzle pressure or gpm
Pump Discharge Pressure Increase:	10 psi
Actual Nozzle Pressure Increase:	5 psi with a gain of 10 gpm

Figure 12–4. Friction loss effect on flow

Figure 12–5. Rounding off the number to determine pressures with this 10 psi increment apparatus gauge makes more sense because it is difficult to be exact.

the discharge end of the gauges with a nozzle. Putting a static pressure on these gauges will show you how close the gauges are to one another (fig. 12–6).

A good practice to follow when conducting flow tests is to have one person designated to read all of the gauges. For some reason I have found that if more than one person is reading gauges, there is a greater likelihood of a gauge being read incorrectly. The gauge reader will have to traverse the fireground to read different gauges, but I think it is worth the effort.

We have already described the difference between an in-line gauge and a pitot gauge. When it comes to measuring nozzle pressures from smoothbore tips, I prefer using the pitot gauge over the in-line gauge. I have found that the in-line gauge connected between the hose and the inlet of the nozzle does not always give an accurate reading compared to the actual nozzle pressure measured at the outlet of the tip. The main reason for this is that the diameter of the waterway at the in-line gauge is usually larger than the tip orifice. This issue can vary from nozzle to nozzle and could be a greater problem with higher flows. A simple flow test proves this theory (fig. 12–7).

Figure 12–6. Check for accuracy between gauges with a static pressure test.

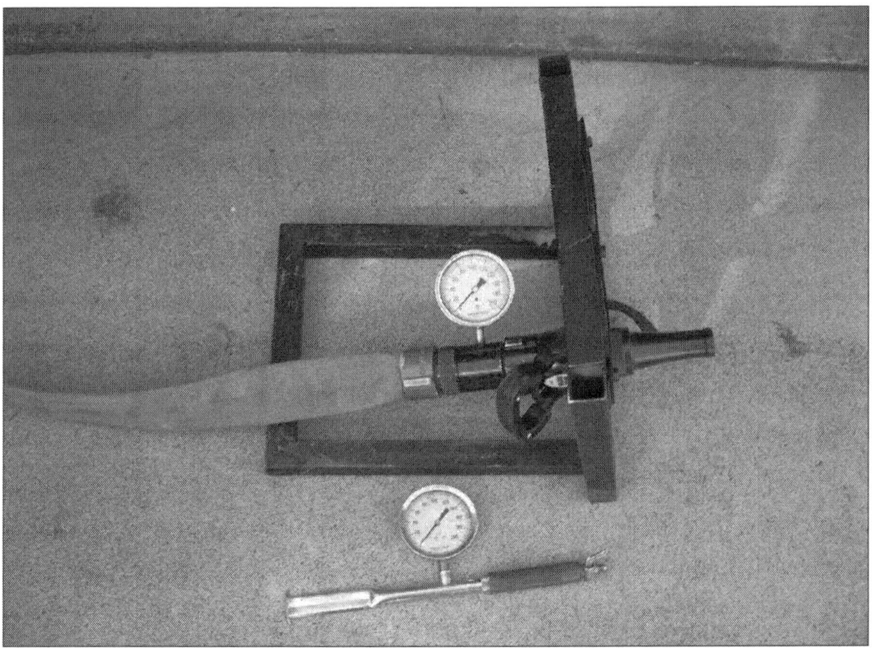

Figure 12–7. Placement of an in-line gauge at the base of a smoothbore nozzle can produce an inaccurate nozzle pressure reading. Always use a handheld pitot gauge.

When the flow passes through the in-line gauge with the 1½" waterway and reduces down to the 15⁄16" tip, it increases in velocity, which in turn raises the pressure (fig. 12–8). You should always use a pitot gauge when measuring a stream from a smoothbore tip. Flow testing combination nozzles, however, is a completely different procedure. First, you cannot use a pitot gauge on a straight stream on a combination nozzle and look up the reading on a corresponding conversion chart. Instead, an in-line gauge should be placed at the inlet of the nozzle to get the most accurate nozzle pressure reading possible. It may not match up perfectly to the actual nozzle pressure, but it is the best we can do. The actual flow will be determined from a flowmeter.

Smoothbore Tip Flow Test	
Tip:	15⁄16-inch
Inline Gauge at the Nozzle Inlet:	50 psi
Nozzle Pressure Measured by a Pitot Gauge:	60 psi

Figure 12–8. Smoothbore tip flow test

When it comes to properly placing the handheld pitot gauge into the stream to measure the pressure, conventional fire instruction requires that the blade of the pitot gauge be centered in the stream, and the distance from the tip to where the blade is positioned should be equal to half the diameter of the tip itself. It is also commonly taught that you must not angle the end of the blade as it is inserted into the stream, keeping it parallel to the stream direction. I have found that the rule about positioning the blade does not seem to hold true. I place the blade in the center of the stream parallel to the stream, but then I rest the blade against my thumb, which braces against the nozzle. See for yourself. First do the placement of the blade as conventionally taught and take a reading. Then place the blade as close to the end of the tip as possible and then once again at a further distance from the tip than usually suggested (fig. 12–9).

Friction Loss in Hose Flow Test

Now let's use the techniques discussed in this section to conduct a flow test to measure friction loss in the hose. It is up to you to decide what length of hose to test. The standard is 100', but when you think about it, we generally use 50' sections on the fireground. The exception is large diameter hose, which is typically

Figure 12–9. The pitot gauge blade is 1" from a 1" tip. Conventional books say that it should be placed at ½". The gauge reading did not change.

used in 100' lengths. You can either test the hose in 50' sections or test 100' sections and divide the friction loss in half.[3]

The following are instructions for conducting flow tests to determine friction loss in fire hose:

1. Determine the flow and use the proper flow measuring device to develop the required flow in the hose. Smoothbore tips require a pitot gauge, and combination nozzles require a flowmeter. These devices should be set up as discussed in this section.
2. You will need two in-line gauges to measure the friction loss. Place one at the beginning of the section to be tested and one at the end of the section to be tested.
3. There should be at least one section of straight hose going into the first gauge and one exiting the second gauge. This will prevent possible turbulence in the gauge itself, which could have a negative effect on the gauge reading. A discharge line of 200' works well for the flow test.
4. Finally, throttle up to the required flow to be tested and read both in-line gauges. Subtracting the lower number from the higher number will give you the friction loss.

Flowmeter

Flowmeters are generally designed to be within ±5% accuracy. Take a look at an example to see how much it can vary. With 185 gpm as the target flow, a reading 5% lower than 185 gpm is 176 gpm, while a reading 5% higher is 194 gpm. When you are conducting flow tests on a combination nozzle, the flowmeter is the only instrument that can be used to achieve the flow. Remember, however, that the number may not be 100% accurate. The best way to assure that the flowmeter is as accurate as possible is to calibrate for the flow you are trying to reach using a smoothbore tip and a handheld pitot gauge. In other words, pitot the stream and recalibrate the flowmeter to match the numbers you get from the pitot gauge. This should be done every time a flow test is conducted (fig. 12–10).[4]

Most manufacturers of flowmeters recommend that there be a straight section of hose 3' to 5' in length going into the flowmeter tube and also one exiting the flowmeter tube. If the flowmeter tube is hooked to the end of the discharge or into the front of the nozzle, turbulence could be created within the flowmeter that could result in an inaccurate reading.

Hydrant Flow Tests

I once talked to a chief about using large diameter hose and its ability to get the most out of a hydrant system. His department used 2½" and 3" hose for its supply lines (fig. 12–11). The chief told me that they did not need large diameter hose because they had an excellent municipal water system with an average pressure of 90 psi. (I assume he meant static pressure.) To his thinking, this indicated a hot hydrant. What he did not realize was that no matter how good a hydrant system is, if you cannot get the water out of the ground through the proper supply line evolution, it will not do much good. The small diameter hose that his department used was actually restricting the capabilities of the hydrant system.[5]

The hydrant flow test sanctioned by *NFPA 291: Recommended Practice for Fire Flow Testing and Marking of Hydrants*, is illustrated in figure 12–12.[6] Two hydrants next to each other on the same main are used for the test. One hydrant is signified as the test hydrant. The static and residual readings are taken from it with no water flowing. The second hydrant is the flow hydrant. A pitot reading is taken from the hydrant port. Three sets of numbers are recorded. They are the static and residual pressures from the test hydrant and the flow pressure from the flow hydrant. These numbers are added to a water flow chart and then graphed to reflect the available water from the main and the available water from the hydrant.

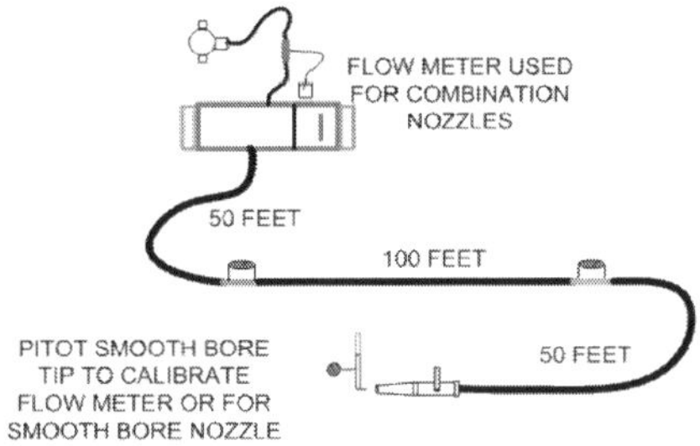

Figure 12–10. Flowmeter used for combination nozzles

Figure 12–11. A short section of 2½" or 3" hose is used before and after the flowmeter tube. This helps alleviate turbulence that might occur if the tube was connected to a discharge or the end of the handline at the nozzle.

The reason we conduct flow tests on hydrants is to give us a good idea what we might get from the hydrant. The flow test will also give us an estimate of the available water in the water main at the point of the hydrant we are using for the flow test. These tests are needed for preplanning as well as for color coding the hydrant to signify the rated flow. Nevertheless, we must be able to get the water out of the ground and into the pumping apparatus to do any good. I believe that an additional flow test should be added to the NFPA standard test. This more realistic test would involve connecting an engine to the hydrant using the actual fireground supply line evolution to obtain a true measure of the water delivery capabilities. I guarantee that the real world flow will be less than the flow test of the hydrant by itself. Ideally this should be a big water flow test. Try to max out the supply line evolution for master stream operations. This could involve a relay

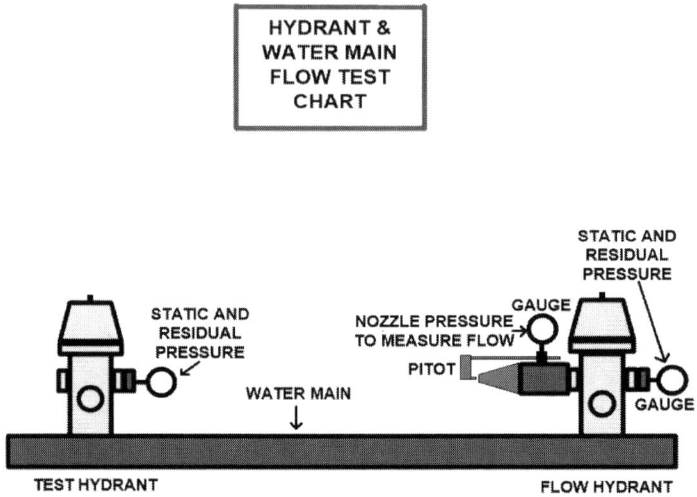

Figure 12–12. A hydrant and water main flow test chart is the standard for testing hydrants and water mains. This system of testing can show a good hydrant but will not show what is available when actually using supply lines.

pump operation, dual supply lines from one hydrant, or dual supply lines from two hydrants. Again, make it realistic according to the standard operating procedures of your department.

Large Flow Evolution

As explained previously, nozzle reaction is the force measured in pounds directly opposite of the stream direction. Nozzle reaction needs to be a part of the handline/nozzle evaluation because of the physical demands it may make on the firefighters. There are mathematical formulas that can accurately calculate most handline/nozzle combinations. The nozzle reaction equation for a smoothbore nozzle is as follows:

$$NR_{Sb} = 1.57 \times D^2 \times NP$$

where

NR_{Sb} is the nozzle reaction for a smoothbore nozzle,

D is the diameter of the nozzle, and

NP is the nozzle pressure.

The formula for nozzle reaction for a combination nozzle is as follows:

$$NR_{Comb} = 0.0505 \times Q^2 \times \sqrt{NP}$$

where

NR_{Comb} is the nozzle reaction for the combination nozzle,

Q^2 is the flow rate, and

$\sqrt{NP}$ is the square root of the nozzle pressure.

Note: This formula will not work for Vindicator nozzles. (See chapter 9 for additional discussion of Vindicator nozzles.)

An alternative to using the formula that is equally accurate is to use a *load scale*, which also measures in pounds. The load scale is designed to measure hanging weight. Have you ever noticed in the hardware store where bulk nails can be purchased by the pound? The unit measuring the weight is a load scale. The scale is anchored from a stationary stand, and the tray holding the nails hangs from the scale on the other end. To measure nozzle reaction with a load scale, the scale is connected between the nozzle and an anchored position. This is a safe operation to use as long as the anchor point and whatever you are using to connect the nozzle to the anchor point are sufficiently strong. I built a fixture that allows the nozzle to be flowed safely while measuring the nozzle reaction. It is attached to the hitch receiver on my pickup truck (fig. 12–13).

Creating the proper fire stream is crucial to a successful outcome at an incident. Proper flow tests will ensure that this can be accomplished. Take the time and do it right.

Figure 12–13. The author built this fixture to measure the nozzle reaction on the Vindicator nozzle using a load scale.

13

DEVELOPING PUMP DISCHARGE PRESSURES

Introduction

Pump operators have one of the most critical jobs within the fire company. They must be able to deliver the troops in a safe and timely manner and provide them with an adequate water flow to suppress the fire as quickly as possible. This involves securing a water supply sufficient to produce the required fire streams to get the job done and to develop the required streams by means of a combination of hose evolutions and proper pump operations.

A pump operator should understand the operation and capabilities of all appliances that involve the movement of water, such as the following:

1. Types of nozzles (smooth bore, automatic, selectable gallonage, fixed gallonage, etc.)
2. The flow ranges of the nozzles
3. Flow ranges, requirements, and capabilities of master stream appliances
4. Hose capabilities, supply, and discharge

There are some things that a pump operator needs to know about physics when it comes to delivering water on the fireground (fig. 13–1). Here is a list of concepts to understand about moving water that will help the pump operator:

- Understanding of friction loss in both hose and plumbing
- Flow rate (gpm)
- Elevation gain or loss (head pressure)
- Proper gauge reading
- Nozzle exit or base pressure

- Nozzle friction loss
- Hose friction loss
- Appliance friction loss
- Static pressure
- Residual pressure

Besides all of the above-mentioned items, the pump operator must also be able to develop the proper discharge pressure to provide the energy to move the water through the hose evolution and discharge it onto the fire.

There are several steps in a hose evolution that require pressure to move the proper amount of water. Back in the old days when buckets were used, the only energy required to move the water was supplied by the people in the bucket brigades. With modern water delivery systems, pressurized water is used to overcome pressure losses from friction in the fire hose, plumbing, and appliances, as well as head pressure loss from elevation and the exit pressure from the stream-producing fire nozzle (also known as nozzle pressure).

There have been several methods used over the years to calculate the required discharge pressure, and some have worked better than others. A standing joke about discharge pressure in the fire service says, "Throttle up until the firefighter and nozzle come off the ground, and then back her off a couple of pounds." Another line of thought is to "pump 150," whether it is for a handline of any size (diameter and length), for master streams, relays, or for building fire protection systems. If

Figure 13–1. A pump operator must understand friction loss and pressure, among other factors.

water comes out the other end, the correct discharge pressure is being developed. Obviously these methods leave much to be desired.

One of the most common methods used for calculating the right discharge pressure is the use of fireground hydraulics. This is a mathematical formula that is based on the size and length of the hose line and the amount of water being delivered, as well as appliance friction loss, elevation gain or loss, building plumbing loss, and nozzle pressure. The old formula $2Q^2 + Q$ was designed for 2½" hose calculations, and it is my understanding that the numbers were actually based on 2½" pipe, not fire hose. A coefficient could then be used in the formula to apply it to other hose diameters.

I know this formula works on paper, but it is not practical or accurate because hose has different characteristics than pipe. Hose curves and sometimes kinks, and it may even expand when it is charged, unlike pipe. It is a proven fact that the same size hose made by different manufacturers can have different friction loss characteristics at a given flow rate (fig. 13–2).

To further complicate matters, there are occasions in which the length, diameter, flow, and other key components of a hose evolution may not be known. The formula cannot be used if you cannot supply all of the numbers. For those who can accurately perform the calculations at the fire scene, my hat goes off to you. It is difficult.

In recent years IFSTA has streamlined the formula to make it more accurate, but in my opinion it is not the best method to use. If there is a good characteristic

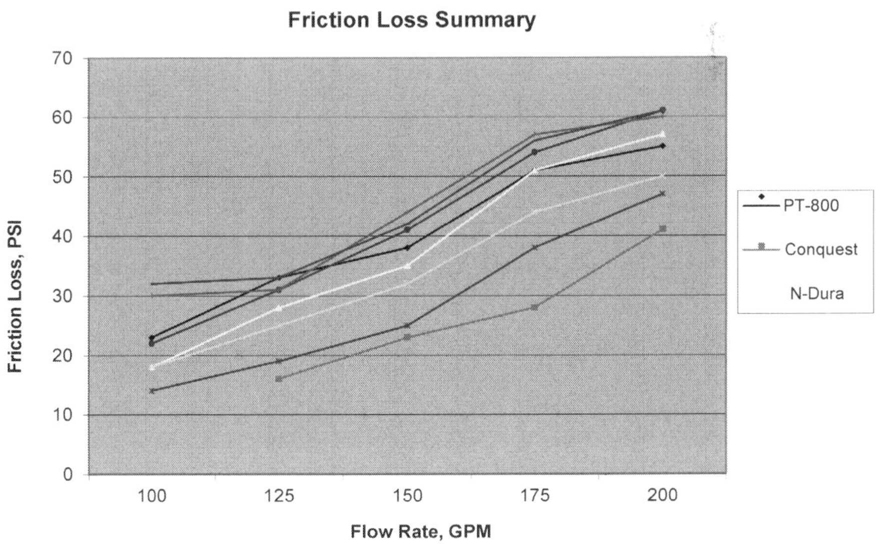

Figure 13–2. This chart shows 1¾" hose manufactured by various companies.

about the calculations for fireground hydraulics, it is that they usually calculate a higher pressure than is actually present.

Flowmeters have become a part of the apparatus industry, generating an electronic readout of the amount of water flowing. Although these instruments can be accurate, most require calibration to remain in good operating order. From what I have seen, most departments do not calibrate their flowmeters. If a flowmeter is to be used, I recommend the type that also has a pressure gauge. This makes for a reliable backup.

Developing Pump Charts

A pump chart is a reference chart for all discharge evolutions that assists the pump operator in developing pump discharge pressures for any hose evolution a department may encounter. Pump charts should be designed to be as easy to read as possible. Make sure the letters and numbers are as large as possible and keep it in a place that is easily accessible and has a light for nighttime use. A predetermined pressure should be calculated for all preconnected evolutions, such as preconnected handlines, fixed master stream appliances, and any evolution that will always be the same (fig. 13–3). The numbers calculated from the flow test should reflect the discharge gauges that are being used by the apparatus. Most discharge pressure gauges read in 10 psi increments. As mentioned previously, I would recommend rounding up to the next highest increment.

Predetermined pump discharge pressures are not the only information that will be handy. There is also a need for a friction loss chart for the various sizes of hose that are used. The reason for this is simple. There are times that lines have to be made up at the fire, which could involve various lengths of hose. By having a friction loss chart the pump operator can add up the length of hose that is being used on a specific line and cross-reference it to the friction loss chart to determine the required pump discharge pressure.

You still need to set up a discharge line and determine the required flow you are testing through the methods just explained. There also must be two in-line pressure gauges connected into the discharge line 100' apart to measure the friction loss. These gauges also need to have 10' of straight hose entering into them for best results. The friction loss is determined by subtracting the lower pressure from the higher pressure on the two gauges.

Finally, it is important to flow test all pumping apparatus in the fleet because their plumbing design and gauge placement could differ, leading to different pump discharge pressures. Every unit should have its own pump chart developed from the actual equipment and evolutions in use.

Figure 13–3. Preconnected lines remain the same, thus allowing for a predetermined discharge pressure.

Handlines

The handline evolution is the most basic discharge evolution used by firefighters. There are two basic parts, the hose and the nozzle. The most common form of this evolution is the preconnected handline.

Preconnects. The method used to determine the correct pump discharge pressure for the preconnected handline is to conduct the basic flow tests previously mentioned. For combination nozzles, a flowmeter is used. For smoothbore nozzles, either a flowmeter or handheld pitot gauge can be use. It is important to flow the lines from the actual preconnected discharge because they are usually smaller in diameter than the 2½" discharges and can have numerous bends in the plumbing design that could affect the required pressure. Additionally, the discharge gauge may be located further away from the end of the discharge plumbing than a standard 2½" discharge gauge.

Handline assembled at the fire. If the handline is developed or assembled at the fire scene and has a different length than the preconnected lines, the final calculation is done by using the friction loss section of the chart. First the length of the handline must be determined. Next the friction loss chart needs to be referenced for the correct size of hose and flow rate (gpm). The friction loss numbers are listed for 100' sections, so simply divide the length of the hose by 100 and multiply the answer by the friction loss listed on the chart to come up with the friction loss for the entire line (table 13–1). It should be noted that the friction loss can be calculated for 50' sections if so desired (fig. 13–4). Finally, add the nozzle pressure based on the actual nozzle that will be used.

Table 13–1. Friction loss in 100' section of hose

Hose Size @ Flow Rate (gpm)	Friction Loss (psi)
1¾" @ 150	25
2½" @ 250	15
2½" @ 300	20
2½" @ 328	25
5" @ 1,000	5

Note: For 100' sections of hose

Figure 13–4. Calculate the friction loss for different hoses.

It is important to understand that an assembled handline can vary slightly from the same length of line on a preconnect. Remember you are working with two different types of discharges, and the pump discharge pressure for the assembled line is developed from numbers on the chart as opposed to using a flowmeter for the preconnect line. Here is an example of what would be on the pump chart:

1¾" preconnect

1¾" 150 gpm handline × 200' with a TFT Automatic 100 psi nozzle: 150 psi PDP.

2½" preconnect

2½" 328 gpm handline × 200' with a 1¼" smoothbore 50 psi nozzle: 100 psi PDP.

2½" handline

As previously discussed in chapter 9, the triple-stack tips are a common smoothbore nozzle used on 2½" handlines (See table 13–2). This nozzle has three tip sizes in a stack, just like a master stream: 1", 1⅛", and 1¼". The easiest way to list

Table 13–2. Pump chart data for a 2½" handline

	100'	200'	300'	400'	500'	600'
2½"	25 psi	50 psi	75 psi	100 psi	125 psi	150 psi

Note: Plus 50 psi for the nozzle pressure for the 1¼"

the pump discharge pressure for a line using this nozzle is to calculate the line using the highest flow tip, which is the 1¼". No matter which tip is being used, always calculate for the 1¼" tip. There is a good chance that the pump operator will not know which tip is being used anyway. If a lower tip size is used, the flow and nozzle pressure will only increase slightly and will still fall within the guidelines for the nozzle's capabilities.

Master Stream Operations

First, I would like to review master stream basics so that the method for determining pump discharge pressures can be better understood. As previously discussed, a master stream is a heavy-caliber stream delivered through a master stream water delivery appliance. A master stream is used when flows surpass 350 gpm and become too difficult to deliver from a handline operation due to nozzle reaction (fig. 13–5). A master stream operation typically produces a high-flow straight stream using a smoothbore tip or a combination nozzle.

There are three types of master stream operations: the fixed master stream, the portable master stream, and the elevated master stream. The following information about master stream smoothbore tips can be found in most fire stream management books currently in circulation. This is information that must be understood in order to establish a base from which to operate.

Smoothbore tips

As previously mentioned, stacked tips are the most common set of smoothbore tips for a master stream. The set consists of four tips connected into one stack, which is then attached to the master stream appliance. The tip sizes and the corresponding flows are based on an 80 psi nozzle pressure (table 13–3).

For master stream appliances capable of flowing 2,000 gpm, the following tip sizes with 80 psi nozzle pressures will apply (table 13–4). Again, these are the most common.

It should be noted that the standard rule concerning smoothbore tip nozzle pressures used with master stream operations is to use a nozzle pressure of 80 psi. According to most fire stream books, nozzle pressures higher than this will create a broken and insufficient stream, as well as cause an unstable operation in the elevated and portable modes.

Now that you have reviewed this information, I would like you to keep an open mind about the following information on smoothbore tip operations. At no time do any of the operations in this text exceed the guidelines set by the manufacturers of the equipment involved.

Portable Master Stream

Master stream operations using smoothbore tips can deliver several different flows, with each potentially having a different pump discharge pressure (fig. 13–6). It is not practical to list all of them on the pump chart. The only pressure that needs to be listed is the pump discharge pressure for the maximum flow operation, with the most common flow being 1,000 gpm with a 2" tip at 80 psi nozzle pressure (table 13–5). This holds true for portable and elevated operations with the supporting discharge evolutions needed to deliver the water to the appliance. When using smoothbore nozzles, the maximum flow pump discharge pressure listed on the pump chart will work with any size tip from 2" down to 1" and will still provide for safe and efficient operations. It is true that the rated flow at the standard smoothbore nozzle pressure (80 psi) will probably be surpassed. However, it will do this while remaining within the required inlet pressure of the appliance (200 psi) and the maximum allowed flow and nozzle reaction.

Figure 13–5. Master streams are used when flows become too difficult to be delivered from a handline.

Table 13–3. Master stream smoothbore tips and corresponding flow at 80 psi

Smoothbore Tip Size	Flow Rate (gpm)
1⅜"	502
1½"	596
1¾"	814
2"	1,063

Table 13–4. High-flow master stream smoothbore tips and corresponding flow at 80 psi

Smoothbore Tip Size	Flow Rate (gpm)
2¼"	1,347
2½"	1,665
2¾"	2,010

Table 13–6 illustrates what would be on the pump chart if all tips were pumped at the same pump discharge pressure for a portable master stream. This chart is for all tip sizes. One pump discharge pressure works for all tips.

High-Pressure Master Stream Smoothbore Tip Operations

The following operations can be used with fixed, elevated, and portable master stream operations. High-pressure smoothbore tip operations have been around for decades. Most firefighters, however, either have not heard about them or have been told that they are counterproductive and dangerous.

The purpose of high-pressure smoothbore tip operations used in conjunction with master streams is to provide the required flow rates to extinguish the fire with a high-velocity stream. The high-velocity stream has improved reach and penetration capabilities while still maintaining efficient stream performance.

The maximum allowed nozzle pressure for this type of operation in a fixed, portable, or elevated master stream operation is based on the following consideration:

Figure 13–6. Master streams using smoothbore tips can deliver different flows depending on pressure.

Table 13–5. Master stream pump discharge pressure

Tip Size	Flow Rate (gpm)	Nozzle Pressure (psi)
1⅜"	589	110
1½"	634	90
1¾"	814	80
2"	1,060	80

Note: Based on a 2" tip flowing 1,060 gpm @ 80 psi nozzle pressure. One pump discharge pressure works for all tips.

1. *Maximum allowed inlet pressure to the master stream appliance.* The three most common manufacturers of the master stream appliance for the municipal fire service are Task Force Tips, Akron Brass, and Elkhart Brass. All have an inlet pressure of 200 psi. Most other master stream appliance manufacturers also require 200 psi inlet pressures. *Never exceed the inlet pressure when performing high-pressure operations.*
2. *Maximum allowed flow for the master stream appliance.*
3. *Maximum allowed nozzle reaction for the master stream appliance.* This is based on a 100 psi combination nozzle at the rated flow of the appliance (table 13–7).

In order to provide the required pressure for the high-pressure operation the pump operator needs to understand the limitations mentioned above and throttle up accordingly. Flow tests will determine what the maximum pressure can be based on the previously mentioned criteria. A pump chart should then be created to show all possible pump operations. The pump operator should plan on throttling up to the maximum allowed predetermined pressure unless something prevents this. (See chapter 9 for an explanation of factors that may stop the operation from going to maximum pressure.) The pump discharge pressure does not

Table 13–6. Portable master stream pump discharge pressures

	100'	200'	300'	400'	500'	600'
5"	100 psi	105 psi	110 psi	115 psi	120 psi	125 psi

necessarily need to be maxed out every time a deck gun operation is placed into service, but this is the goal the pump operator will try to reach. Often tips up to 1½" can be pumped up to 200 psi pump discharge pressure.

High-Flow, High-Velocity Operations

Smoothbore nozzles

The following smoothbore tip sizes with their corresponding nozzle pressures and flows have successfully produced high-velocity/high-flow streams using a master stream (table 13–8). The maximum nozzle reaction for a 1,250 gpm appliance is 631 lbs. For a 2,000 gpm appliance, the maximum nozzle reaction is 1,010 lbs. (See chapter 9 for a more detailed discussion.)

An example of what would be listed on the pump chart is shown in table 13–9.

Combination nozzles

The same concept described with the smoothbore tips in regard to pumping the highest flow also applies to combination nozzles.

Selectable gallonage nozzle. The pump discharge pressure should be established for the highest flow in the flow range of the nozzle. If a lower flow in the flow range is selected, there will only be a slightly elevated nozzle pressure, nozzle reaction, and flow rate, which will all remain well within the operational operations.

Fixed-gallonage nozzle. The pump discharge pressure should be established for the rated flow of the nozzle. If the resulting flow is less than what the nozzle is rated at, the reach of the fire stream will decline. Whether or not the reach is sufficient will depend on the actual fireground scenario.

Table 13–7. Nozzle reactions for master stream appliances with 100 psi combination nozzle

Flow Rate (gpm)	Nozzle Reaction (lbs reactive force)
1,250	631
1,500	758
2,000	2,010

Automatic nozzle. For a large flow operation, the pump discharge pressure should be established for the highest rated flow. If the flow rate is not achievable, the automatic nozzle will adjust its pressure to whatever flow is being delivered while maintaining optimum stream performance.

Elevated Master Stream

Table 13–8. Smoothbore nozzle tips and corresponding pressures and flow rates

1,250 gpm Appliance			
Tip Size	Nozzle Pressure (psi)	Flow Rate (gpm)	Nozzle Reaction (lbs reactive force)
1⅜"	175	743	519
1½"	175	884	618
1¾"	120	996	577
2"	100	1,189	628
2,000 gpm Appliance			
Tip Size	Nozzle Pressure (psi)	Flow Rate (gpm)	Nozzle Reaction (lbs reactive force)
2"	140	1,407	879
2"	150	1,455	942
2"	160	1,500	1,005

Table 13–9. Example pump chart with tip size and corresponding pump discharge pressure (PDP) for a fixed master stream with smoothbore tips

Fixed Master Stream with Smoothbore Tips	
1⅜", 1½", and 1¾" tips	200 psi PDP
2" tip	140 psi PDP

Elevated master streams are similar to fixed master streams with one exception. They are either mounted on aerial ladders, aerial platforms, or TeleSqurts. A *TeleSqurt* is a small elevated stream mounted on an engine. It is usually only 50' in height and can be equipped with a ladder. If the unit with the elevated master stream has a pump, then the same flow tests used for the fixed master stream will also work for the elevated stream. Always follow the chart on the pedestal in regard to flow versus angle versus height to ensure safe operations.

There are also elevated streams on truck companies that do not have pumps. These require a different method for determining pump discharge pressure, but it can easily be determined with a flow test. Most truck companies carry the feeder hose that is going to the ladder of the truck company to feed the waterway going up to the nozzle. It is usually a 100' section of hose. Regardless of how long it is, the friction loss of hose can be calculated from a hose friction loss chart as the first step. The second part of the problem involves conducting a flow test with a flowmeter or handheld pitot gauge to determine nozzle pressure for the required flow. Next, simply add the two together to determine the entire pump discharge pressure.

The third step is the most involved part of this flow test. For these purposes, elevation is accounted for with a basic formula of 0.5 psi/1', which must be added to the pump discharge pressure. If a flow test needs to be performed to determine the required flow through the full extension of the ladder itself, you may decide to go for the maximum and develop the pump discharge pressure at the fully extended mode and use that throughout the extension of the ladder (fig. 13–7). The key information to determine here is if the ladder is extended lower than the full position but is using the full position pump discharge pressure, how much will the flow rate and nozzle pressure rise? It may be similar to using the stacked tips or the combination nozzle where you will see a small rise. The results would still be within the recommendations of the manufacturer for the use of the equipment (i.e., the strength of the ladder, nozzle, and appliance). On the other hand, it may be a situation in which lowering the height of the nozzle to a certain point causes the operation to become unsafe. This should be noted on the pump chart and *never exceeded*. There is a possibility of coming up with a lower flow that will work at that height, but the maximum flow should not be used.

Building Fire Protection Systems

A building fire protection system is a plumbing system within a structure designed to deliver water for fire suppression within its confines (fig. 13–8). The water supply is either provided by the system water supply, which is plumbed into a municipal water source, or by an FD pumping apparatus connected to a water source. The final discharge of the water is accomplished through strategically installed sprinkler heads and/or fire department attack lines via standpipe hose connections.

The following information are the basic codes set forth by *NFPA 14, Standard for the Installation of Standpipe and Hose Systems,* 2010 edition, outlining pressure limitations for building fire protection systems that have a direct effect on how the fire department delivers water within the structure.[1]

(a)

(b)

Figure 13–7. To see the required flow, a flow test is needed with the full extension of the ladder.

- System hydrostatically tested at no less than 200 psi.
- For a system pressure of 150 psi or higher, the hydrostatic test will be no less than 50 psi higher than the system pressure.
- The maximum allowed system pressure is no higher than 350 psi.
- Express standpipes can exceed 350 psi.
- Buildings with a single pump as well as express standpipes for multiple pumps can exceed 350 psi to meet the standards on tall buildings.

As discussed previously, there are two basic sets of standards established by NFPA that affect water delivery. The first standard was in effect until 1993, when it was replaced by the second standard. Again, structures built in compliance with the pre-1993 standards were not required to upgrade to the second standard. (See chapter 6 for more details.)

Figure 13–8. This photo shows a tandem pump operation using LDH and 2½" discharges to go into the FDC.

Sprinkler systems

Pump operations for building sprinkler systems are basic (fig. 13–9). The system itself is rated at 200 psi and the pump discharge pressure for the operation is 150 psi. It is important to remember that the 150 psi needs to be maintained. With the possibility of additional sprinkler heads activating after the 150 psi pressure is established, the engine speed may need to be increased via the throttle to maintain the pressure. The pump chart should have a note under the heading "Sprinkler Systems" with instructions to maintain 150 psi.

Multistory building fire protection system with no building pump

Low-rise building standpipe systems without system pumps, whether they are standpipes only or combination systems, can be found in low-rise structures (fig. 13–10). These systems can be either wet or dry. If the system is wet, it is supplied from municipal water, which does not necessarily provide enough pressure to support handlines. If it is dry, the FD pumping apparatus needs to provide all the pressure. It is the responsibility of the FD pumping apparatus to increase the system pressure to the required pressure to support the handlines deployed by the firefighters.

Throughout the years the pump discharge pressures for systems like this have always been calculated using fireground hydraulics. The requirements include the following:

- 25 psi for the standpipe
- 5 psi for each floor
- Pressure requirements for the handline, including friction loss for the hose and the nozzle pressure
- Friction loss in the lines supplying the FDC from the pumper

The following questions must be answered for the formula to be usable:

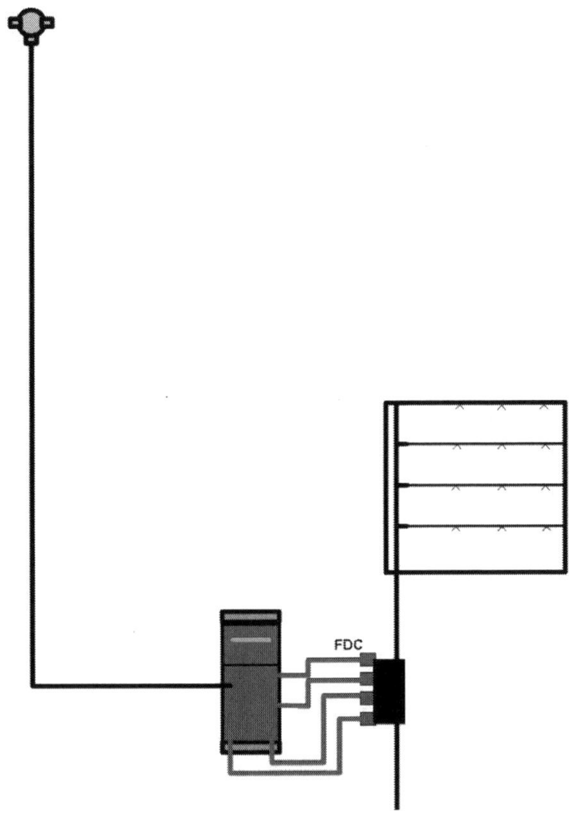

Figure 13–9. Basic pump operations for building systems

- Does the pump operator know what floor the handline or handlines are connected on?

- Does the pump operator know how many handlines are being used?
- Does the pump operator know the makeup of the handline (length, diameter, nozzle pressure, and gpm)?

In the real world of firefighting, there is a good chance that the pump operator will not have all the answers. Additionally, formulas can generate answers that will be hard to find on a standard pumper 10 psi increment discharge gauge. The alternative is to pump a maximum pressure, 200 psi. Here is why. The 200 psi number is based on the maximum allowed system pressure. There is no guessing. The 200 psi gives the firefighters maximum pressure, which in turn will provide

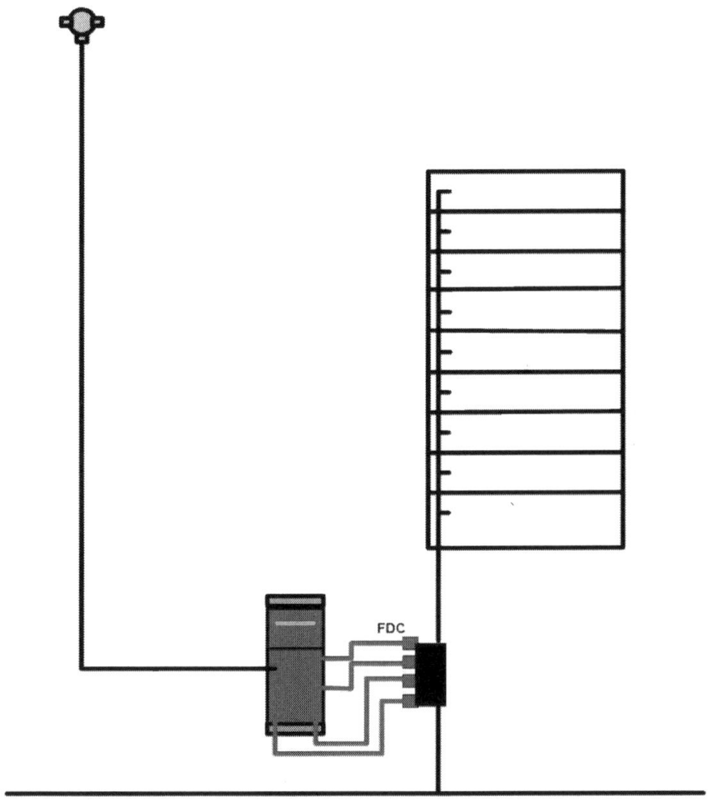

Figure 13–10. Low-rise structures have standpipe systems with no system pump.

maximum flow. If the pressure and flow are excessive, the firefighter can control them by gating the standpipe outlet valve.

There is an excellent chance that the pump discharge pressure developed from the hydraulic formula will be close to 200 psi anyway. An example evolution will demonstrate how this might work.

Example evolution. This scenario involves a three-story building with fire on the third floor. The evolution will use 150' of 1¾" hose with a 100 psi automatic nozzle flowing 150 gpm.

Calculations

10 psi for elevation

150 psi for the 1¾" handline

25 psi for the standpipe

Negligible for the lines supplying the FDC

Total: 185 psi

The pump chart should have a note under the heading "Standpipe System with No Building Pump" with instructions to maintain 200 psi.

High-rise building fire protection system with a building pump

If the fire department needs to pump the system in a high-rise building, it will have to match the building pump pressure and possibly exceed it (fig. 13–11).[2]

As explained in chapter 6, high-rise building fire protection systems are equipped with check valves on the discharge side of the building pump. Because of these one-way check valves, if the pump is running but not producing the necessary pressure or flow rate, the only way the fire department can take control of the system is to exceed the building pump pressure. Even if the building pump is not running, there will still be system head pressure up against the check valve that will have to be overcome by the engine companies.

If the building pump fails to provide the required pressure, the fire department pumper must still pump the system as though the fire is on the roof (at the highest pressure). The reason for this is simple. A building fire pump does not have the capability of controlling its engine pressure over the wide pressure range that would be needed to provide the proper pressure to all the floors in a multistory building. Therefore, the building pump is designed to pump to the highest point in the building. Because of the potential for high pump pressures in the system, as explained in chapter 6, different types of pressure-regulating devices are incorporated into the system, including pressure relief valves. These valves are designed to reduce the building pressure to the proper pressure on the floor where it is in service.

Two pump discharge pressures

After all lines are connected, the system pressure should be determined and the engine pressure of the pumper supplying the system should be set at 50 psi below the system pressure. Setting the FD pumper engine pressure 50 psi below the building system pressure will allow the building pump to stay running and do the work as long as it can. A pressure drop in the system of more than 50 psi is a good indication of a substantial amount of water flowing, which in turn is a good indication of a drop in residual pressure that could affect the flow to the handlines. It might also be an indication of pump failure. It is at this point the FD pumpers should take over the building system.

Once the fire department pumper is set up and pumping its required engine pressure, it should be determined whether or not the pumper is flowing water

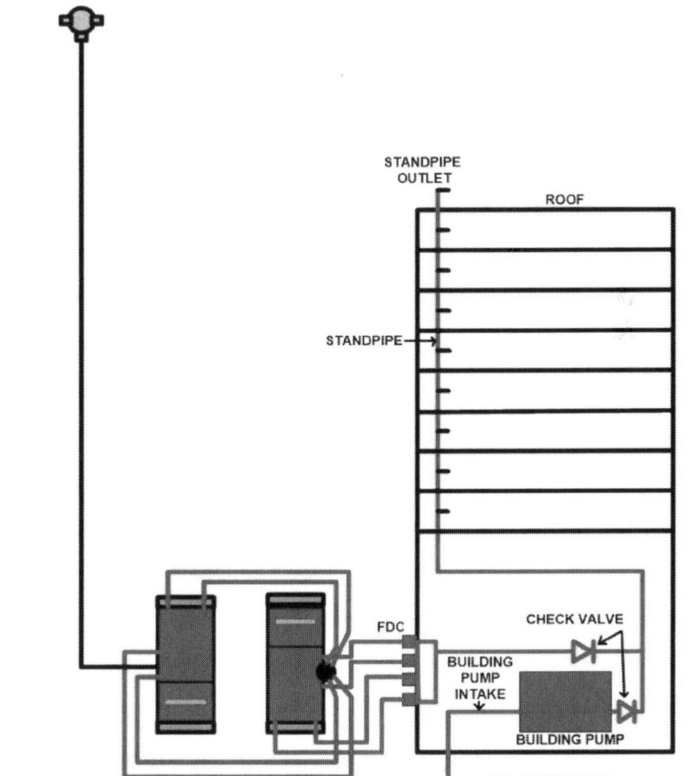

Figure 13–11. In many high-rise buildings, the fire department needs to match the building pump pressure or exceed it.

into the system. When the pumper starts flowing water, whether initially after it goes into operation or at any time during the incident, or if the firefighters inside request more pressure, it is important to immediately *pump the system pressure.* (For more information see chapter 6 and table 6–2.)

Relay Pump Operations

A relay pump operation is a supply line evolution that is required when there is not enough pressure from the water source to move the required volume of water to the fire through the supply hose (fig. 13–12).[3] This operation requires a pumper to be connected into the supply line evolution at the water source to boost the pressure in the supply line to move the water to the pumper at the fire. This holds true for both short-distance and long-distance operations. A relay pump operation comes in handy to support existing pump operations on the fireground with additional water. Developing the required pump discharge pressure is simple. When the units are hooked up and ready to flow water, the line is charged under low pressure, similar to charging a supply line from a hydrant. Once the line is charged, the pumper at the water source throttles up to an initial pressure to give the receiving pumper enough water to get started. I like to start at about 120 psi. Once the pump operator at the receiving end starts flowing, he or she can communicate to the pump operator at the water source and advise if more pressure is needed. Here is an example of what the pump chart should look like.

Figure 13–12. A pumper connected to a supply line at the water source can boost enough pressure to move water to the pumper.

Sample chart for relay pump operation

Start with 120 psi pressure. Table 13–10 shows how far a single pumper can move specific amounts of water. Based on this data, a decision can be made concerning how many pumpers will be needed in a relay. This can also be placed on a pump chart.

Table 13–10. Distance vs. flow rate for a single pumper

4" LDH	
Hose Length	Flow Rate (gpm)
500'	1,500
600'	1,250
1,100'	1,000

CONCLUSION

I hope this book has been helpful in expanding your knowledge of large flow water delivery. As you can see, there is more to a large volume fire attack than just lobbing water. All personnel involved with fire suppression should have a thorough knowledge of the setup and operation of large flow water delivery. This training should start in rookie school and continue indefinitely. Quarterly multicompany drills should be conducted to keep firefighters informed of any new concepts, as well as to review the existing concepts. Since large flow fires occur so infrequently, firefighters may lack firsthand experience in fighting them.

It is common knowledge that the greatest amount of dollar loss annually due to fire in the United States is caused by only a small percentage of the fires. Successful firefighting at these fire scenes will likely require high flows. You must be on your game.

NOTES

Introduction

1. Paul Shapiro, "Big Paulie's Words of Wisdom on Movin' the Big Water," FireFlowTechnology.com, http://fireflowtechnology.com/big-paulies-words-of-wisdom-on-movin-the-big-water/.
2. National Fire Protection Association, "Care, Use, Inspection, Service Testing, and Replacement of Fire Hose," ch. 4 in *NFPA 1962: Standard for the Care, Use, Inspection, Service Testing, and Replacement of Fire Hose, Couplings, Nozzles, and Fire Hose Appliances*, 2018 ed. (Quincy, MA: NFPA, 2017), https://www.nfpa.org/codes-and-standards/all-codes-and-standards/list-of-codes-and-standards/detail?code=1962.

1. Water Supply Fundamentals

1. Wikipedia, s.v. "Pressure Head," https://en.wikipedia.org/wiki/Pressure_head.
2. National Fire Protection Association, *NFPA 1901: Standard for Automotive Fire Apparatus*, 2016 ed. (Quincy, MA: NFPA, 2015), https://www.nfpa.org/codes-and-standards/all-codes-and-standards/list-of-codes-and-standards/detail?code=1901.
3. Paul Shapiro, "Back to Basics, Laying a Large Diameter Hose Supply Line," FireFlowTechnology.com (February 28/March 10, 2014), http://fireflowtechnology.com/back-to-basics-laying-a-large-diameter-hose-supply-line/.
4. "What Are Custom Length Hoses Used For?" RawhideFireHose.com, https://rawhidefirehose.com/blog/custom-length-hoses-used/.
5. Wikipedia, s.v. "Cavitation," https://en.wikipedia.org/wiki/Cavitation.
6. Paul Shapiro, "The Hose Mule Hose Retrieval System," *Fire Apparatus & Emergency Equipment* 20, no. 7 (July 9, 2015), http://www.fireapparatusmagazine.com/articles/print/volume-20/issue-7/features/the-hose-mule-hose-retrieval-system.html.
7. National Fire Protection Association, *NFPA 1500: Standard on Fire Department Occupational Safety, Health, and Wellness Program*, 2018 ed. (Quincy, MA: NFPA, 2017), https://www.nfpa.org/codes-and-standards/all-codes-and-standards/list-of-codes-and-standards/detail?code=1500.

2. Pumping from the Hydrant

1. "Pumping from the Hydrant," FireFlowTechnology.com (February 28/March 10, 2014), http://fireflowtechnology.com/pumping-from-the-hydrant/.
2. Paul Spurgeon, "Every Pump Operator's Basic Equation," FireEngineeringUniversity.com (October 2012), https://www.fireengineering.com/wp-content/uploads/content/dam/fe/online-articles/documents/FEU/SpurgeonOct12.pdf.

3. "Big Streams Require a Proactive Water Supply," FireFlowTechnology.com (April 13, 2015), http://fireflowtechnology.com/big-streams-requires-a-proactive-water-supply/.
4. Paul Shapiro, "Supplemental Water Supply Operations," Fire Rescue 10, no. 11 (November 1, 2015), https://firerescuemagazine.firefighternation.com/2015/11/01/supplemental-water-supply-operations/#gref.
5. Shapiro, "Supplemental Water Supply Operations."
6. Wikipedia, s.v. "Water Hammer," https://en.wikipedia.org/wiki/Water_hammer.
7. "Net Pump Discharge Pressure (NPDP)," FirefighterMedic.com (July 29, 2013), http://www.firefightermedic.com/wiki/net-pump-discharge-pressure-npdp/.
8. Shapiro, "Supplemental Water Supply Operations."
9. Shapiro, "Supplemental Water Supply Operations."
10. Shapiro, "Supplemental Water Supply Operations."
11. Shapiro, "Supplemental Water Supply Operations."

3. Standard Units Used for Big Flow Operations—You Already Own Them

1. Paul Shapiro, "Apparatus for Big Flow Operations—You Already Own Them, Part 1," Fire Apparatus & Emergency Equipment 23, no. 5 (May 1, 2018), https://www.fireapparatusmagazine.com/2018/05/01/apparatus-for-big-flow-operations-you-already-own-them-part-1/.
2. Wikipedia, s.v. "Ferrara Fire Apparatus," https://en.wikipedia.org/wiki/Ferrara_Fire_Apparatus.
3. John Lehoczky, Jr., "New York's Super Pumper System," *Fire Engineering* 10, no. 118 (October 1, 1965), https://www.fireengineering.com/1965/10/01/324233/new-yor k-s-super-pumper-system/.
4. Shapiro, "Apparatus for Big Flow Operations—You Already Own Them, Part 1."
5. Selena K. Chin, Peter B. Sunderland, and Grunde Jomaas, "Firefighter Nozzle Reaction," *Fire Technology* 53, no. 5 (May 10, 2017), https://backend.orbit.dtu.dk/ws/portalfiles/portal/137621163/Firefighter_Nozzle_Reaction_Chin_Jomaas_Sunderland_Fire_Technology_.pdf.
6. Paul Shapiro, "Apparatus for Big Flow Operations—You Already Own Them, Part 2," *Fire Apparatus & Emergency Equipment* 23, no. 8 (August 1, 2018), https://www.fireapparatusmagazine.com/2018/08/01/apparatus-for-big-flow-operations-you-already-own-them-part-2/.
7. Shapiro, "Apparatus for Big Flow Operations—You Already Own Them, Part 2."
8. Shapiro, "Apparatus for Big Flow Operations—You Already Own Them, Part 1."
9. Shapiro, "Apparatus for Big Flow Operations—You Already Own Them, Part 2."

5. Delivering 2,000 gpm from an Elevated Stream

1. Wikipedia, s.v. "Quint (Fire Apparatus)," https://en.wikipedia.org/wiki/Quint_(fire_apparatus).

2. Paul Shapiro, "Delivering 2,000 gpm from an Elevated Stream," *Fire Rescue* 10, no. 9 (September 1, 2015), http://www.firerescuemagazine.com/articles/print/volume-10/issue-9/firefighting-operations/delivering-2-000-gpm-from-an-elevated-stream.html.
3. Shapiro, "Delivering 2,000 gpm from an Elevated Stream."
4. Shapiro, "Delivering 2,000 gpm from an Elevated Stream."
5. Shapiro, "Delivering 2,000 gpm from an Elevated Stream."
6. Shapiro, "Delivering 2,000 gpm from an Elevated Stream."

6. Water Delivery Operations for Fires in High-Rise Structures

1. National Fire Protection Association, *NFPA 14: Standard for the Installation of Standpipe and Hose Systems*, 2019 ed. (Quincy, MA: NFPA, 2018), https://www.nfpa.org/codes-and-standards/all-codes-and-standards/list-of-codes-and-standards/detail?code=14.
2. Paul Shapiro, "Pump Operations for High-Rise Building Fire Protection Systems," *Fire Engineering* (August 29, 2012), http://www.fireengineering.com/articles/2012/08/pump-operations-for-high-rise-building-fire-protection-systems.html.
3. Paul Shapiro, "High-Rise, Big Water, Part 1," *Fire Rescue* 11, no. 5 (May 1, 2016), http://www.firerescuemagazine.com/articles/print/volume-11/issue-5/firefighting-operations/high-rise-big-water-part-1.html.
4. Shapiro, "Pump Operations for High-Rise Building Fire Protection Systems."
5. National Fire Protection Association, *NFPA 1961: Standard on Fire Hose*, 2020 ed. (Quincy, MA: NFPA, 2019), https://www.nfpa.org/codes-and-standards/all-codes-and-standards/list-of-codes-and-standards/detail?code=1961; and National Fire Protection Association, *NFPA 1962: Standard for the Care, Use, Inspection, Replacement, and Service Testing of Fire Hose, Couplings, Nozzles, and Fire Hose Appliances*, 2018 ed. (Quincy, MA: NFPA, 2017), https://www.nfpa.org/codes-and-standards/all-codes-and-standards/list-of-codes-and-standards/detail?code=1962.
6. Shapiro, "Pump Operations for High-Rise Building Fire Protection Systems."
7. Shapiro, "High-Rise, Big Water, Part 1."
8. Shapiro, "Pump Operations for High-Rise Building Fire Protection Systems."
9. Shapiro, "High-Rise, Big Water, Part 1."
10. Shapiro, "Pump Operations for High-Rise Building Fire Protection Systems."
11. Shapiro, "High-Rise, Big Water, Part 2."
12. NFPA, NFPA 14: Standard for the Installation of Standpipe and Hose Systems.
13. Shapiro, "High-Rise, Big Water, Part 2."
14. Shapiro, "Task Force Tips Gated Siamese Manifold Finding Its Way into High-Rise Pump Operations," *Fire Apparatus & Emergency Equipment* (September 29, 2015), http://www.fireapparatusmagazine.com/articles/2015/09/task-force-tips-gated-siamese-manifold-finding-its-way-into-high-rise-pump-operations.html.

7. Water Management Officer

1. Paul Shapiro, "Water Management Officer: Taking Charge of the Water," *Fire Engineering* (June 12, 2012), http://www.fireengineering.com/articles/2012/06/water-management-officer-taking-charge-of-the-water.html.
2. Shapiro, "Water Management Officer: Taking Charge of the Water."
3. Shapiro, "Water Management Officer: Taking Charge of the Water."
4. Shapiro, "Water Management Officer: Taking Charge of the Water."

8. Comparing 4" and 5" Large Diameter Hose

1. Paul Shapiro, "Comparing Four- and Five-Inch Large Diameter Hose," *Fire Apparatus & Emergency Equipment*, 19, no. 7 (July 1, 2014), http://www.fireapparatusmagazine.com/articles/print/volume-19/issue-7/features/comparing-four-and-five-inch-large-diameter-hose.html.
2. Shapiro, "Comparing Four- and Five-Inch Large Diameter Hose."
3. Shapiro, "Comparing Four- and Five-Inch Large Diameter Hose."
4. Shapiro, "Comparing Four- and Five-Inch Large Diameter Hose."
5. Shapiro, "Comparing Four- and Five-Inch Large Diameter Hose."
6. Shapiro, "Comparing Four- and Five-Inch Large Diameter Hose."
7. Shapiro, "Comparing Four- and Five-Inch Large Diameter Hose."
8. Shapiro, "Comparing Four- and Five-Inch Large Diameter Hose."
9. Shapiro, "Comparing Four- and Five-Inch Large Diameter Hose."

10. The 2½" Handline

1. Paul Shapiro, "Big Paulie: The Monster Handline," FireFlowTechnology.com (April 27, 2015), http://fireflowtechnology.com/big-paulie-the-monster-handline/.
2. Shapiro, "Big Paulie: The Monster Handline."
3. Paul Shapiro, "A Small Line with a Big Punch," *Fire Rescue* 11, no. 4 (April 1, 2016), https://firerescuemagazine.firefighternation.com/2016/04/01/a-small-line-with-a-big-punch/.
4. Bill Nemick, "Piercing Nozzle Tactic for Attic Fires," *Fire Engineering* 170, no. 10 (October 1, 2017), https://www.fireengineering.com/articles/print/volume-170/issue-10/features/piercing-nozzle-tactic-for-attic-fires.html.
5. Nemick, "Piercing Nozzle Tactic for Attic Fires."
6. Shapiro, "A Small Line with a Big Punch."
7. Paul Shapiro, "The Design of the High-Rise Handline," *Fire Rescue* 11, no. 7 (November 1, 2016), https://firerescuemagazine.firefighternation.com/2016/07/01/the-design-of-the-high-rise-handline/?topic=20089.
8. Shapiro, "A Small Line with a Big Punch."
9. Shapiro, "A Small Line with a Big Punch."

10. Paul Shapiro, "Two-Inch Hose: The Lightweight Preconnected Little Big Line," *Fire Apparatus & Emergency Equipment* (June 2014), 20–21, http://digital.fireapparatusmagazine.com/fireapparatus/201406?pg=22#pg22.
11. Shapiro, "Two-Inch Hose: The Lightweight Preconnected Little Big Line."
12. National Fire Protection Association, *NFPA 1962: Standard for the Care, Use, Inspection, Service Testing, and Replacement of Fire Hose, Couplings, Nozzles, and Fire Hose Appliances* 2018 ed. (Quincy, MA: NFPA, 2017), https://www.nfpa.org/codes-and-standards/all-codes-and-standards/list-of-codes-and-standards/detail?code=1962.
13. Shapiro, "Two-Inch Hose: The Lightweight Preconnected Little Big Line."
14. Shapiro, "The Design of the High-Rise Handline."
15. National Fire Protection Association, *NFPA 14: Standard for the Installation of Standpipe and Hose Systems*, 2019 ed. (Quincy, MA: NFPA, 2018), https://www.nfpa.org/codes-and-standards/all-codes-and-standards/list-of-codes-and-standards/detail?code=14.

11. The Big Hit Offensive Concept

1. Paul Shapiro, "The Big Hit Offensive Concept," FireFlowTechnology.com (June 1, 2017), http://fireflowtechnology.com/the-big-hit-offensive-concept/.
2. Shapiro, "The Big Hit Offensive Concept."
3. Howard Sykes and Thomas Sturtevant, *Fire Apparatus Operator: Pumper* (Clifton Park, NY: Delmar, Cengage Learning, 2012), 338.
4. National Fire Protection Association, *NFPA 1964: Standard for Spray Nozzles*, 2018 ed. (Quincy, MA: NFPA, 2017), https://www.nfpa.org/codes-and-standards/all-codes-and-standards/list-of-codes-and-standards/detail?code=1964.
5. Shapiro, "The Big Hit Offensive Concept."
6. Shapiro, "The Big Hit Offensive Concept."
7. Shapiro, "The Big Hit Offensive Concept."

12. Measuring Water Flow for Fire Suppression

1. Paul Shapiro, "Measuring Water Flow for Fire Suppression," *Fire Apparatus & Emergency Equipment* 21, no. 4 (April 15, 2016), https://www.fireapparatusmagazine.com/2016/04/15/measuring-water-flow-for-fire-suppression/.
2. Shapiro, "Measuring Water Flow for Fire Suppression."
3. Shapiro, "Measuring Water Flow for Fire Suppression."
4. Shapiro, "Measuring Water Flow for Fire Suppression,"
5. Shapiro, "Measuring Water Flow for Fire Suppression,"
6. National Fire Protection Association, *NFPA 291: Recommended Practice for Fire Flow Testing and Marking of Hydrants*, 2019 ed. (Quincy, MA: NFPA, 2018), https://www.nfpa.org/codes-and-standards/all-codes-and-standards/list-of-codes-and-standards/detail?code=291.

13. Developing Pump Discharge Pressures

1. National Fire Protection Association, *NFPA 14, Standard for the Installation of Standpipe and Hose Systems*, 2010 ed. (Quincy, MA: NFPA, 2009), https://www.nfpa.org/codes-and-standards/all-codes-and-standards/list-of-codes-and-standards/detail?code=14.
2. Paul Shapiro, "Pump Operations for High-Rise Building Fire Protections Systems," *Fire Engineering* (August 29, 2012), http://www.fireengineering.com/articles/2012/08/pump-operations-for-high-rise-building-fire-protection-systems.html.
3. Paul Shapiro, "Maximizing the Water Supply on a Large Flow Fire with Supplemental Pump Operations," FireFlowTechnology.com (April 13, 2015), accessible from http://fireflowtechnology.com/articles/.

INDEX

F

O

W